REVISE AQA GCSE
Science A

NEL

REVISION GUIDE
Higher

Series Consultant: Harry Smith

Authors: Sue Kearsey, Nigel Saunders and Peter Ellis

THE REVISE AQA SERIES
Available in print or online

Online editions for all titles in the Revise AQA series are available Spring 2013.

Presented on our ActiveLearn platform, you can view the full book and customise it by adding notes, comments and weblinks.

Print editions

Science A Revision Guide Higher	9781447942146
Science A Revision Workbook Higher	9781447942153

Online editions

Science A Revision Guide Higher	9781447942207
Science A Revision Workbook Higher	9781447942238

Print and online editions are also available for Science (Foundation), Additional Science (Higher and Foundation) and Further Additional.

This Revision Guide is designed to complement your classroom and home learning, and to help prepare you for the exam. It does not include all the content and skills needed for the complete course. It is designed to work in combination with Pearson's main AQA GCSE Science 2011 Series.

D1355491

To find out more visit:
www.pearsonschools.co.uk/aqagcsesciencerevision

ALWAYS LEARNING

PEARSON

Contents

A small bit of small print

AQA publishes Sample Assessment Material and the Specification on its website. This is the official content and this book should be used in conjunction with it. The questions in Now try this have been written to help you practise every topic in the book. Remember: the real exam questions may not look like this.

Target grades

Target grade ranges are quoted in this book for some of the questions. Students targeting this grade range should be aiming to get most of the marks available. Students targeting a higher grade should be aiming to get all of the marks available.

A healthy diet

A healthy diet contains the right amounts and proportions of NUTRIENTS and energy that the body needs to stay healthy.

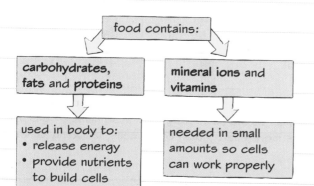

food contains:

| carbohydrates, fats and proteins |
| mineral ions and vitamins |

used in body to:
• release energy
• provide nutrients to build cells

needed in small amounts so cells can work properly

A person who does not eat a balanced diet may become MALNOURISHED.

So what we eat can affect our health.

Possible effects include:
• being very overweight or underweight
• **deficiency diseases** caused by too little of a nutrient
• conditions such as **Type 2 diabetes**.

Anything with this Spec Skills sticker is helping you to *apply* your knowledge. You can find out more about these stickers on page 95.

AQA SKILL Evaluate Page 95

Worked example

target D-C

AQA SKILL Evaluate Page 95

A scientific study showed that for a 10 g increase in wholegrain fibre that a person ate each day, there was a 10% decrease in their risk of developing bowel cancer. The scientists concluded that people should eat lots of wholegrain foods every day to help them stay healthy.

Do the results of this study support this conclusion? Give a reason for your answer.

(2 marks)

The results do support this conclusion, because the more fibre the person eats the lower their risk of getting bowel cancer.

You will not be expected to know the effect of fibre in the diet, but you will be expected to answer questions like this about the effect of food on health from data that you are given. You will be given data to work with.

Now try this

target D-C
target B-A*

1 What is meant by the term malnourished? *(1 mark)*

2 Iron is a mineral needed in the diet. A diet deficient in iron can cause a disease called anaemia. The table shows the recommended iron intake for different age groups of females.

Age in years	4–8	9–13	14–18	19–50	51+
Recommended intake in mg/day	10	8	15	18	8

(a) Which age group is most at risk of anaemia? Give a reason for your answer. *(1 mark)*

(b) Pregnant women are more at risk of anaemia than women who are not pregnant. Explain why.

(2 marks)

Controlling mass

Metabolic rate

METABOLIC RATE is the rate at which all the chemical reactions are carried out in the body.

Metabolic rate is affected by many factors:

- how much muscle you have
- how much exercise you do
- some INHERITED FACTORS (factors in your genes).

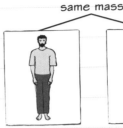

same mass

bigger muscles so less fat

smaller muscles so more fat

A man usually has a higher metabolic rate than a woman of the same mass, because muscle cells use lots more energy than other cells, including fat cells.

The body gains energy by eating food. Energy is EXPENDED (used) during exercise. The balance between energy taken in and energy expended affects your mass.

more energy expended in exercise than gained in food

more energy gained in food than expended in exercise

body loses mass   body gains mass

Remember: the correct scientific word for your 'weight' is mass.

Slimming products and programmes

Slimming programmes and slimming food products claim to help you lose mass. Slimming programmes include guidance about what to eat and how to exercise, as well as providing support to help you keep to targets.

You may be asked to interpret data on claims for slimming products or programmes in your exam.

Worked example

target D-C

AQA SKILL
Evaluate
Page 95

The chart shows the mean loss of mass in people on three different slimming programmes compared with a group who only exercised.

The makers of Programme B claim theirs is the best slimming programme. Does the graph support this claim? Give reasons for your answer. *(2 marks)*

Loss of mass after 12 weeks, and then after one year, was greater on the other two programmes. So it is not the best of these programmes for losing mass.

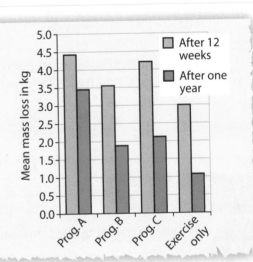

Mean mass loss in kg

☐ After 12 weeks
☐ After one year

Prog. A   Prog. B   Prog. C   Exercise only

Now try this

target D-C

1 Explain why eating less and exercising more can help you lose mass. *(3 marks)*

2 Compare the slimming and exercise-only data in the chart above. Explain which is the best way to lose mass. *(3 marks)*

Lifestyle and disease

LIFESTYLE is the way we live, including what we eat and what we do, which affects how active we are. Lifestyle factors can harm health and lead to disease.

Exercise and disease

A person who exercises regularly is more likely to stay healthy than a person who doesn't.

benefits of exercise

better weight control ⟹ better health

Inherited factors

INHERITED FACTORS (genes) are not lifestyle factors, but they can affect health. For example, some people have genes that give them a higher blood cholesterol level than other people who eat the same diet.

When we study the effect of lifestyle factors on disease, we must compare people with similar inherited factors so that we get results that can be trusted.

Worked example target D-C

In a study, 522 overweight people of similar age who were at risk of developing diabetes were separated into two groups. One group was given advice on how to live more healthily, the other received no advice (control group). After 4 years 11% of the advised group and 23% of the control group had developed diabetes.

The scientists concluded that living healthily reduces the risk of developing diabetes. Do the data support the scientists' conclusion? Give reasons for your answer.   *(3 marks)*

The data do support the conclusion because 11% of the advised group developed diabetes. This is much less than the 23% of the control group. We can trust the study because it contained a large group (522) of people, of similar age, who all started with the same problems.

You need to comment on how valid the results are as well as what the conclusion is. Validity increases with larger numbers and by having a control group to make it a fair test.

Now try this

 1 Explain why exercise can increase a person's chances of staying healthy.
(2 marks)

 2 A study of over 43 000 American adult men showed that overweight men who were physically fit had the same risk of diseases such as Type 2 diabetes as men who were not overweight, but that unfit overweight men had a much higher risk of these diseases. A simple conclusion from this study is that being fit is important for health as well as not being overweight.
Evaluate this conclusion.   *(3 marks)*

EXAM ALERT!

For an 'evaluate' answer, use your knowledge and the information given to consider evidence for and against, and then draw a suitable conclusion from your arguments.

Students have struggled with questions like this in recent exams – **be prepared!**

Pathogens and infection

Microorganisms that cause disease are called PATHOGENS. Pathogens include some bacteria and viruses. When a few pathogens INFECT us (get inside our bodies) they can reproduce very rapidly. Large numbers of pathogens can make us ill.

Bacteria are much smaller than our cells.

bacterium

Bacteria may release TOXINS (poisons) that make us feel ill. Some types of bacteria invade and destroy body cells.

Viruses are much smaller than bacteria.

virus

Viruses take over a body cell's DNA, causing the cell to make TOXINS or causing damage when new viruses are

Semmelweis

Ignaz Semmelweis was a doctor in the mid-1800s who wondered why many women died of infection soon after childbirth.

 People didn't know about microorganisms then, so they didn't know what caused infection.

| Semmelweis realised that doctors might transfer infection between patients on their hands. | → | He insisted that doctors wash their hands before examining each patient. | → | Death rates fell rapidly in wards where doctors washed their hands. |

Worked example

target **D-C**

AQA SKILL Interpret Page 95

Semmelweis stated that cleaning hands before treating patients could reduce the number of infections in hospitals. The data shown are from a modern hospital where infections can sometimes still be a problem. In 2007 the hospital started a campaign to remind staff to wash their hands before treating patients. Is Semmelweis's statement supported by the data?   *(2 marks)*

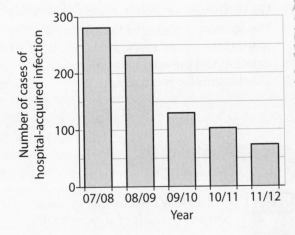

The number of infections caught by patients has decreased each year since 2007. This partly supports Semmelweis's idea, but not completely, otherwise the number of infections would have dropped suddenly.

Now try this

 target **D-C**

1 Explain why Semmelweis's idea of making doctors wash their hands prevented the spread of some infections.   *(2 marks)*

2 Explain why it takes some time after we are infected before we feel ill.   *(2 marks)*

 target **B-A***

3 Suggest **two** reasons why the results in the graph above did not show a single drop in infection rate. Explain your answers.   *(2 marks)*

The immune system

The body has different ways of protecting itself against pathogens.

The IMMUNE SYSTEM helps to protect the body against pathogens. WHITE BLOOD CELLS are part of the immune system.

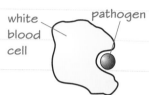

Some white blood cells flow round (INGEST) pathogens and destroy them.

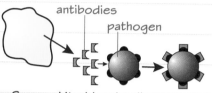

Some white blood cells produce chemical ANTIBODIES that attach to pathogens and destroy them.

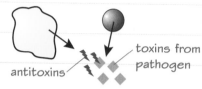

Some white blood cells produce ANTITOXINS that destroy toxins made by pathogens.

Antibodies

The antibodies produced by a white blood cell are SPECIFIC for one particular kind of pathogen. This means they can only destroy that kind of pathogen. They cannot destroy another kind of bacterium or virus.

Immunity

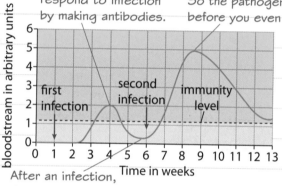

White blood cells respond to infection by making antibodies.

The white blood cells release more antibodies more quickly. So the pathogens are killed before you even feel ill.

Amount of antibody in bloodstream in arbitrary units

first infection

second infection

immunity level

Time in weeks

After an infection, the immune system remembers how to attack that pathogen.

Worked example

target D-C

Measles is caused by a virus. Explain why you cannot suffer from measles more than once. *(2 marks)*

Your immune system remembers how to respond to the pathogen that causes measles and will destroy another infection of measles so quickly that you do not fall ill.

Now try this

target D-C

1 You have recovered from an infection that causes measles. Explain why you could fall ill from a different infection, such as the pathogen that causes chickenpox. *(2 marks)*

target B-A*

2 Some children cannot be vaccinated against childhood infections because their immune systems are weakened by other illness. Explain how these children are unlikely to catch the childhood infections if most healthy children have been vaccinated. *(2 marks)*

Immunisation

IMMUNISATION means to make someone immune to a disease.

Vaccination

VACCINATION is a way of making someone immune to a disease by giving them a VACCINE. The MMR VACCINE is given to children to make them immune to measles, mumps and rubella for the rest of their lives.

| A vaccine contains a small amount of a dead or inactive form of a pathogen. | → | The vaccine causes white blood cells to make antibodies, in the same way they would if the body was infected by live pathogens. | → | If the live pathogen infects you later, your immune system remembers how to destroy it, and responds quickly so you don't fall ill. You are immune. |

Mutations of pathogens

| A **mutation** produces a new strain of a pathogen. | → | Antibodies against the old strain may not recognise the new strain. | → | People who were vaccinated against the old strain are not immune to the new strain. | → | The new strain may spread rapidly, causing an **epidemic** or **pandemic** disease. |

An epidemic is when many people catch a disease at the same time. A pandemic is when many people in many places have the same disease at the same time.

Worked example

Use the information in the graph to explain the importance of vaccination in preventing disease.

(3 marks)

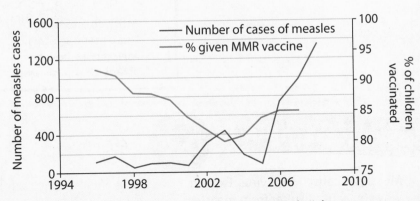

The graph shows that the percentage of children vaccinated with MMR fell from over 90% in 1996 to around 80% in 2003. From 2005, the number of cases of measles rose rapidly. This suggests that a decrease in the proportion of children vaccinated increases the risk of being infected.

Now try this

1 A parent might decide not to get their child vaccinated using the MMR vaccine.

    (a) Suggest **one** reason for this decision.   *(1 mark)*

    (b) Suggest **one** reason why it would be better if their child is vaccinated.   *(1 mark)*

2 The graph above suggests a link between vaccination and infection rate, but there may be other factors that increased the infection rate. Suggest **one** other reason the infection rate might have increased.   *(1 mark)*

Treating diseases

Some medicines, such as painkillers, only treat SYMPTOMS of disease. They do not kill pathogens. Other medicines help you by killing the pathogens.

Using antibiotics

- ANTIBIOTICS are medicines that kill bacterial pathogens inside the body.
- Specific bacteria are only killed by a specific antibiotic, so the correct antibiotic must be used.
- Deaths from bacterial diseases have greatly decreased where antibiotics are used.

A symptom is the result of disease, such as feeling pain or having a high temperature. It is not the cause of the disease.

Penicillin is an example of an antibiotic.

Antibiotic resistance

| Mutation can produce new antibiotic-resistant strains of bacteria. | → | When the antibiotic is used, non-resistant bacteria die, but resistant bacteria survive and reproduce. | → | The population of resistant bacteria increases. Infections can only be treated with a new antibiotic. | → | If there is no new antibiotic to control the infection, it may spread rapidly, causing an epidemic or pandemic. |

This is an example of natural selection.

Remember that antibiotics do not cause bacteria to become resistant, nor does failing to finish a course of antibiotics.

Many bacteria, such as MRSA, have developed antibiotic resistance.

Problems with viruses

Viruses reproduce inside the cells of another organism and damage the cells. However, antibiotics do not affect viruses.

Drugs that kill viruses may also harm human cells, so viral diseases can be hard to treat.

Worked example

Explain why antibiotics are no longer used to treat non-serious infections. *(3 marks)*

Using antibiotics kills non-resistant strains of bacteria, leaving resistant strains to reproduce and increase in population size. This increases the risk of people being infected by strains against which the antibiotic will no longer work. Non-serious infections will be controlled by the body's immune system.

Now try this

D-C

1 Which statement explains why antibiotic-resistant bacteria are becoming common. *(1 mark)*
   A Antibiotics mutate the bacteria so they become resistant.
   B Overuse of antibiotics means only resistant bacteria survive.
   C Antibiotics used on viruses make them mutate bacteria.

B-A*

2 European doctors are concerned about the discovery of a strain of tuberculosis bacterium that is resistant to several antibiotics and has killed many patients in recent years. Suggest how this strain developed and explain why doctors are worried. *(4 marks)*

Cultures

The action of disinfectants and antibiotics can be studied using cultures of microorganisms. Other microorganisms from the air and surfaces can easily contaminate cultures when they are being prepared. Several techniques can help prevent this.

 Sterilising dishes and culture media

Microorganisms grow in a culture medium. More than one medium are referred to as **media**.

STERILISATION kills microorganisms.

- Petri dishes can be sterilised by autoclaving or heating to a high temperature.
- Culture media (the substance that the microorganisms grow on, such as nutrient agar) are sterilised by heating to a high temperature.

 Sterilising inoculating loops

The loop is sterilised in a hot flame and then cooled, before using it to transfer microorganisms to the growth medium.

 Sealing Petri dishes

The lid is secured to the dish with adhesive tape to stop microorganisms from the air getting in.

 Worked example **target D-C**

Cultures of microorganisms should be incubated at a maximum temperature of 25 °C in school and college laboratories. Explain why.
(2 marks)

Temperatures higher than this encourage rapid growth in bacteria. This includes pathogenic bacteria that are harmful to humans.

When microorganisms are grown in industry, they are cultured at higher temperatures than 25 °C because they are not handled by people. The microorganisms grow more quickly at these higher temperatures, but there is no risk that people will get infected by the bacteria.

 Now try this

target D-C 1 Give **three** ways in which cultures of microorganisms can easily be contaminated with other microorganisms during preparation.
(3 marks)

 target B-A* 2 Describe fully how you would prepare a culture of microorganisms so that you could investigate the action of a disinfectant on the microorganisms.
(4 marks)

Biology six mark question 1

There will be one 6 mark question on your exam paper which will be marked for *quality of written communication as well as scientific knowledge*. This means that you need to apply your scientific knowledge, present your answer in a logical and organised way, and make sure that your spelling, grammar and punctuation are as good as you can make them.

Worked example

There are many vaccinations that can be given to young children. A few years ago the number of children being vaccinated with the MMR vaccine fell because one doctor had suggested that the vaccine could cause children to develop autism. Explain why it is important for parents to get their children vaccinated.   *(6 marks)*

Vaccination makes you immune to the disease. So if the child is vaccinated, it won't ever be able to catch the disease.

If the child caught the disease, they would have a much greater chance of being very ill. So it's much better to be vaccinated. If most children are vaccinated then the disease won't be able to spread easily.

EXAM ALERT!

Always plan what you are going to write for the six mark questions. You are given credit for a well-organised answer.

Students have struggled with questions like this in recent exams – **be prepared!**

Command words

The COMMAND WORD in this question is 'explain'. This means that you need say WHAT is happening and WHY.

This part of the answer is not complete because it doesn't describe how vaccination produces immunity – by stimulating the immune system to produce antibodies, which leads to the development of immunity.

This part of the answer could be even better if it said more clearly that although there is a very small risk of harm from a vaccination, there is a much greater risk of harm if you catch the disease.

Now try this

Explain how a balanced diet and regular exercise can help you stay healthy.   *(6 marks)*

Receptors

Humans react to their surroundings using their nervous system.

| stimulus
change in
surroundings | → | nervous system
detects stimuli and
coordinates response | → | response
change in organism's behaviour
as a result of stimulus |

Note the spelling:
one **stimulus**, two
or more **stimuli**.

Cells that detect stimuli are called RECEPTORS. Receptor cells in different organs of
the body respond to different kinds of stimuli.

Receptor organs	Contain cells sensitive to ...
eye	• light
ear	• change in position (helps us balance) • sound
nose	• chemicals in air (smell)
tongue	• chemicals in solids and liquids (taste)
skin (different cells for each stimulus)	• touch, pressure, pain, temperature

An **organ** contains
different groups
of cells that work
together for a
particular purpose.

Worked example D-C

The diagram shows a light receptor cell from a human eye,
labelled to show three cell structures that are also found
in most other animal cells. Write down the names of cell
structures A, B and C.   *(3 marks)*

A: cytoplasm
B: nucleus
C: cell membrane

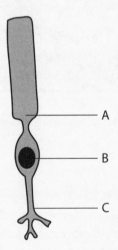

Now try this

D-C

1 A girl puts her hand on a sharp pin and quickly moves her hand away again.
   (a) Identify the stimulus and response in this example.   *(2 marks)*
   (b) What kind of receptors in the girl's body received the stimulus?   *(1 mark)*

B-A*

2 Write down definitions for the following terms:
   (a) stimulus, **(b)** receptor, **(c)** response, **(d)** sense organ.   *(4 marks)*

10

Responses

Nerves contain nerve cells (NEURONES) that connect the CENTRAL NERVOUS SYSTEM (spinal cord and brain) to receptors and EFFECTORS. The brain coordinates the response by different effectors.

REFLEX ACTIONS are automatic responses that are very fast and do not involve conscious thought. They involve only two or three neurones of different types.

(1) A receptor cell responds to a stimulus by producing an electrical IMPULSE.

(2) The impulse passes from the receptor along a SENSORY NEURONE to the central nervous system.

(3) The impulse passes from the sensory neurone to a RELAY NEURONE in the central nervous system.

(4) The impulse passes from the relay neurone to a MOTOR NEURONE.

(5) The impulse passes along the motor neurone to the EFFECTOR.

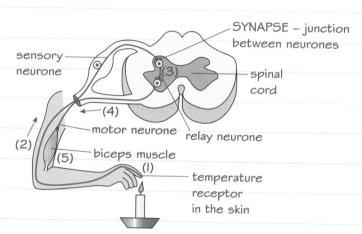

SYNAPSE – junction between neurones

sensory neurone

spinal cord

(4)

motor neurone   relay neurone

(2)

(5)   biceps muscle

(1)

temperature receptor in the skin

Effectors are:
• muscles that respond by contracting
• glands that respond by **secreting** (releasing) chemicals, e.g. hormones.

Worked example   target D-C

Describe the role of synapses in the nervous system.   *(4 marks)*

A synapse is a junction between two neurones. The electrical impulse cannot cross this gap. When the electrical impulse arrives at the synapse, it triggers the release of a chemical into the gap in the synapse between the two neurones. The chemical crosses the gap and triggers a new electrical impulse in the next neurone.

EXAM ALERT!

Remember that nerves and neurones are not the same. Nerves are collections of neurones (nerve cells). Also, neurones carry electrical **impulses**, not messages.

Students have struggled with questions like this in recent exams – **be prepared!**

Now try this

1 Describe the role of the three different neurones in a reflex arc.   *(3 marks)*

2 Explain why simple reflex actions are rapid and why this is important.   *(3 marks)*

Controlling internal conditions

Internal conditions in the body are controlled so that the body works properly.

Water content

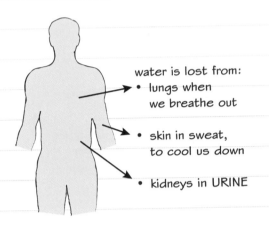

water is lost from:
- lungs when we breathe out
- skin in sweat, to cool us down
- kidneys in URINE

Ion content

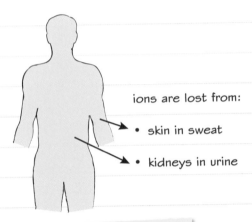

ions are lost from:
- skin in sweat
- kidneys in urine

Water and ions are taken into the body in food and drink.

Worked example target D-C

Body temperature and blood sugar concentration are two other conditions that are controlled in the body. Explain why they are controlled. *(2 marks)*

Body temperature is controlled to keep the body at a temperature at which enzymes work best.

Blood sugar concentration is controlled so that cells are provided with a constant supply of energy.

Hormones

Many processes in the body are controlled by chemicals called HORMONES.

| Hormones are **secreted** (released) by a **gland** into the blood. | → | Hormones are transported around the body in the blood. | → | Hormones cause a response from a **target organ**. |

Now try this

 target D-C

1 Name a hormone in the human body, the gland where it is produced and its target organ. *(3 marks)*

 target B-A*

2 Explain why it is important that body temperature is controlled. *(2 marks)*

The menstrual cycle

The MENSTRUAL CYCLE in women is controlled by several hormones. They cause the MATURATION (ripening) and release of eggs from the ovaries, and cause changes in the thickness of the womb lining.

Three hormones control the release of an egg each month.

> You won't be expected to know details of the menstrual cycle, just the roles of the hormones.

> OESTROGEN inhibits (prevents) further production of FSH so no more eggs mature this month

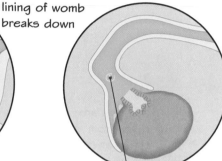

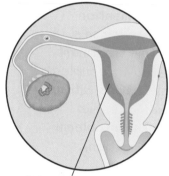

lining of womb breaks down

ovary containing eggs

lining of womb gets thicker

FOLLICLE-STIMULATING HORMONE (FSH)
- causes eggs in ovaries to mature
- stimulates ovaries to produce hormones including oestrogen

LUTEINISING HORMONE (LH)
- causes an ovary to release an egg

> FSH and LH are secreted by the pituitary gland below the brain.
> Oestrogen is secreted by the ovaries.

Oral contraceptives

ORAL CONTRACEPTIVES are 'birth control' pills, as they help to prevent pregnancy. They contain hormones that inhibit the release of FSH, which means:

- no eggs mature
- so no eggs are ready to be released from an ovary.

Oral contraceptives may contain the hormones OESTROGEN and PROGESTERONE.

> Remember that FSH causes eggs to mature, it does not cause eggs to be produced.

Worked example

The first contraceptive pills contained much higher doses of oestrogen than modern pills, and some modern pills contain only progesterone. State why the amounts and type of hormone were changed.   *(1 mark)*

Large amounts of oestrogen caused side effects in the women who took them. The newer pills cause fewer side effects but are still effective.

Now try this

1 Why are the hormones progesterone and oestrogen used in oral contraceptive pills?   *(3 marks)*

2 Explain how hormones interact to control the release of an egg from an ovary during the menstrual cycle.   *(3 marks)*

Increasing fertility

FERTILITY is the ability to have children.
- Contraceptive pills reduce fertility.
- Fertility drugs can increase fertility.

Fertility drugs

FERTILITY DRUGS contain the hormones FSH and LH. The drugs can help women who produce too little FSH by stimulating eggs to mature and then be released.

IVF (in vitro fertilisation)

IVF is fertilisation outside a woman's body. This treatment is offered to couples who are having difficulty conceiving a child (i.e. having problems with fertilisation).

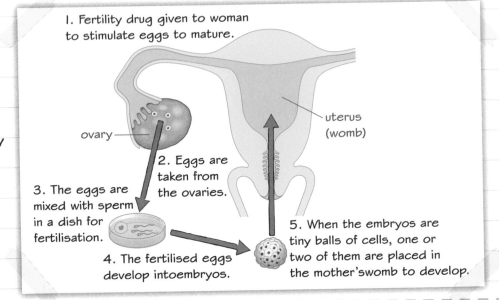

1. Fertility drug given to woman to stimulate eggs to mature.

2. Eggs are taken from the ovaries.

3. The eggs are mixed with sperm in a dish for fertilisation.

4. The fertilised eggs develop into embryos.

5. When the embryos are tiny balls of cells, one or two of them are placed in the mother's womb to develop.

ovary — uterus (womb)

Worked example

B-A*

The table shows the proportion of IVF treatments carried out in the UK in 2008 that resulted in a baby for mothers of different ages.

Age of mother	<35	35–37	38–39	>40
Proportion of successful treatments	33.1%	27.2%	19.3%	10.7%

Compare the benefits and problems that may arise from using IVF in fertility treatments. (3 marks)

IVF helps some couples have a baby when they might not otherwise be able to. However, the woman has to go through treatment which might be uncomfortable. As a woman's age increases, the chance of having a baby decreases.

Now try this

1 Describe how IVF can make it possible for a couple to have a baby when the woman doesn't normally release matured eggs from her ovaries. (4 marks)

2 The box opposite shows some information on the use of the contraceptive pill. Use the information to evaluate the benefits and problems that may arise from using contraceptive pills to control fertility. (3 marks)

Around $\frac{1}{3}$ of UK women of reproductive age take a contraceptive pill. If used properly, the pill is 100% effective against pregnancy. Studies over 40 years show that the pill reduces the risk of many cancers. If the woman smokes heavily or is obese, taking the pill greatly increases the risk of thrombosis (blood clot). The increase in risk of thrombosis for women who don't smoke or are not overweight is very small, far less than the risk of thrombosis during pregnancy.

14

Plant responses

shoots grow:
- towards light
- against gravity

roots grow:
- towards moisture
- in the direction of gravity

Plants respond to changes in light, moisture and gravity by changing how they grow.
- Growth in response to light is called PHOTOTROPISM.
- Growth in response to gravity is called GRAVITROPISM.

Gravitropism is sometimes called **geotropism**. You can use either word in the exam.

Plant hormones

AUXIN is the plant hormone that controls phototropism and gravitropism. Auxin is produced in the shoot or root tip, then moves away from the tip to where it affects cells.
- Auxin causes plant shoot cells to ELONGATE (get longer) more rapidly.
- Auxin INHIBITS (reduces) elongation in plant root cells.

If different cells in a root or shoot contain different amounts of auxin, this causes UNEQUAL GROWTH RATES. This will change the direction of growth of the whole shoot or root.

Auxin and gravitropism

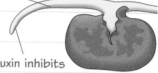

gravity causes auxin to move to lower part of root

more auxin inhibits elongation of lower cells so root curves downwards

Worked example   D-C

This shoot is growing towards light as a result of phototropism. Explain what is happening to auxin at point A, and to the cells at point B.
(3 marks)

A
shaded side of shoot

light

B

lit side of shoot

At A, light is causing the auxin to move to the shaded side of the shoot. At B, cells on the shaded side contain more auxin than cells on the lit side. So the cells of the shaded side grow longer and the shoot grows towards the light.

EXAM ALERT!

Remember: auxin has a different effect on different cells.

Students have struggled with questions like this in recent exams – **be prepared!**

Now try this

 D-C

1 Compare the responses of plant shoot cells and root cells to auxin.
(2 marks)

2 Describe how unequal distribution of auxin causes a horizontal root to start growing downwards. *(3 marks)*

 B-A*

3 Charles Darwin investigated the growth of plant shoots. He placed a dark cap over the tip of a shoot, then grew the plant in one-sided light. Describe what you think the result of this experiment was and explain your answer. *(3 marks)*

Plant hormones

Plant growth hormones can be used in:
- AGRICULTURE (growing crop plants in fields)
- HORTICULTURE (growing flowers, fruit and vegetable plants).

Weedkillers (herbicides)

WEEDS are plants that grow where we don't want them. Plant hormones are used in SELECTIVE WEEDKILLERS. These kill broad-leaved weeds in grass lawns and in crops that have narrow leaves, such as wheat. Killing the weeds in crops reduces competition for water and nutrients in the soil.

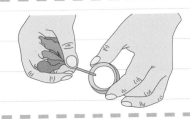
The hormones in the weedkiller affect the weeds so that they die, but don't harm the grass.

Rooting hormones

Plant hormones are used in ROOTING POWDER. If the stem of a plant cutting is dipped in the powder, the stem will develop roots more quickly.

AQA SKILL
Explain
Page 95

The diagram shows two germinating maize seeds. The seeds have been kept in identical conditions, apart from the amount of hormone in the water they were given.

Explain whether you think rooting powder produces better roots.   (3 marks)

grown in pure water   grown in water containing hormone

maize seeds

roots

The root of the seed grown with hormone is longer than the one with no hormone. I think that the hormone increases rate of root growth, because that was the only difference in the way the seeds were kept.

EXAM ALERT!

Although there are differences in the shoots of these seeds, the question does not ask for this. So don't include this in your answer.

Students have struggled with questions like this in recent exams – **be prepared!**

Now try this

target D-C

1 Explain why plant cuttings dipped in rooting powder develop roots more quickly than if rooting powder is not used.   (2 marks)

2 Explain why farmers use weedkillers on their crops.   (2 marks)

target B-A*

3 Populations of seed-eating farmland birds have decreased greatly since farmers started using selective weedkillers on their crops. Suggest **one** reason for the relationship and describe **one** way that bird populations could be protected.   (2 marks)

New drugs

DRUGS are chemicals that affect how the body works. Scientists are continually developing new drugs. New MEDICAL DRUGS must be extensively tested before doctors can PRESCRIBE them to patients. There are several stages of testing.

1 In the laboratory

Drugs are tested on:

cultures of cells    cultures of tissues

animals

These MODELS help predict how the drugs may behave in the human body.

2 Clinical trials: stage 1

healthy volunteer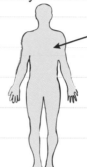

very small DOSE of drug

To check that the drug is not TOXIC (harmful).

Medical drugs are prescribed (given) by doctors to help patients who are ill.

3 Clinical trials: stage 2

 different doses of drug

patient with the disease that the new drug is developed for

To test EFFICACY (whether it works) and to find the OPTIMUM dose (the dose that works best).

Placebos

In some trials, some of the patients are given a PLACEBO. This is something that looks like the drug but doesn't contain the drug.

Results from patients who took the placebo are compared with results from patients who had the drug. Any difference suggests how well the drug works.

Worked example   B-A*

Explain how a double-blind trial is carried out. (3 marks)

Double-blind because two people (doctor and patient) don't know.

In a double-blind trial, neither the doctor nor the patient know if the patient is receiving the placebo or drug until after the trial is complete. This is because the patient might be happier about taking the drug rather than the placebo, and this can affect the results. Also, if the doctor does not know what the patient took, they will be more objective about the results.

Now try this

D-C 1 Give **three** reasons why new drugs need to be fully tested before doctors can prescribe them to patients. (3 marks)

B-A* 2 Suggest **one** advantage and **one** disadvantage of using animals as models for testing new drugs rather than direct testing on humans. (2 marks)

Thalidomide and statins

Thalidomide and statins are medical drugs.

Using thalidomide

THALIDOMIDE caused problems when it was first prescribed by doctors.

| developed as a sleeping pill | → | also used to control morning sickness in pregnant women | → | caused severe limb abnormalities in babies born to many of the women who took the drug | → | thalidomide was then banned |

 not tested on pregnant women

Thalidomide is now used successfully to treat diseases such as **leprosy**.

As a result of the problems with thalidomide, drug testing has been made more thorough.

Worked example   **target D-C**

AQA SKILL Evaluate Page 95

A major study was carried out on the effectiveness of statins on patients with cardiovascular disease. 2223 patients were given a placebo and 2221 were treated with statins. After about 5 years there were 30% fewer deaths from heart disease in the statin group compared with the control group.

Should the National Health Service give permission to doctors to prescribe statins? Give the reasons for your answer.   *(2 marks)*

There were 30% fewer deaths over 5 years when people used the statins compared to people not taking the drug. This is a big saving in lives and cost to the Health Service. So it seems a good idea for the doctors to prescribe statins to protect against cardiovascular disease.

Statins can reduce the risk of cardiovascular disease in people with high blood cholesterol. Cardiovascular diseases are diseases of the heart and circulatory system.

Now try this

target D-C

1   Explain why the effect of thalidomide on babies was not expected.   *(2 marks)*

target B-A*

2   Taking statins may encourage the development of diabetes. A large double-blind trial tested for this by dividing people into one group with risk factors for diabetes and one group with no risk factors. Within each group, people were given either a placebo or a statin. The table shows results from the trial over 5 years.

Group	% difference in statin takers who developed condition compared with those in group taking placebo	
	cardiovascular disease	diabetes
no diabetes risk	52% reduction	no change
diabetes risk	39% reduction	28% increase

Use the results to recommend whether statins should be prescribed to patients in each group. Explain your answers.   *(4 marks)*

Recreational drugs

A RECREATIONAL DRUG is a drug that people use for its effect rather than for medical purposes.

RECREATIONAL DRUGS

LEGAL drugs
(allowed by law)
e.g.
• alcohol
• nicotine (in tobacco)

ILLEGAL drugs
(forbidden by law)
e.g.
• ecstasy
• cannabis
• heroin

Smoking tobacco and drinking too much alcohol can lead to many health problems.

These drugs can harm the heart and circulatory system.

Cannabis

Cannabis is a drug that is usually burned and the smoke breathed in. The smoke contains many chemicals. Some of these chemicals may cause MENTAL ILLNESS in some of the people who use cannabis, particularly if they are regular users while still teenagers.

Addiction

Some drugs, such as heroin and cocaine, are very ADDICTIVE. This means they change chemical processes in the body, and make the person DEPENDENT on taking the drug. They may need more and more of the drug to have the same effect. If the person stops taking the drug, they suffer distressing WITHDRAWAL SYMPTOMS.

Worked example

 D-C

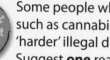
Suggest
Page 95

Some people who take illegal 'soft' drugs, such as cannabis, progress to taking 'harder' illegal drugs such as heroin. Suggest **one** reason for this. *(1 mark)*

They may get used to the effect of the soft drugs and want to try something that has a bigger effect.

Remember that drugs alter chemical processes in the body. They do not cause withdrawal symptoms directly.

Other possible answers are:
• They may like taking risks and the harder drugs are bigger risks.
• Buying illegal soft drugs may give them access to illegal hard drugs, which will tempt them to try the hard drugs.

Now try this

1 Explain why heroin users find it difficult to stop taking the drug. *(2 marks)*

2 A study of over 1000 people from birth to age 38 showed that those who smoked cannabis on a regular basis in their teens and twenties performed less well in IQ tests at age 38 than those who did not take the drug. This was after taking into account other factors that might have an effect, such as alcohol or smoking tobacco. The more cannabis a person took, the greater the decrease. Evaluate the investigation. *(3 marks)*

Drugs and health

Drugs can be evaluated by comparing their beneficial and harmful effects, to decide which has the greater impact.

The impact of drugs

Many more people are treated in hospital for the effects of legal drugs, such as tobacco or alcohol, than for the use of illegal drugs. This is because there are many more people who smoke or misuse alcohol than people who take illegal drugs. So there will be more who suffer harm from tobacco and alcohol than suffer harm from illegal drugs.

This doesn't mean that illegal drugs are safer to use than legal (prescribed and non-prescribed) drugs. The **proportion** of illegal drug users who suffer harm is greater than the proportion of legal drug users who suffer harm.

Worked example **target D-C**

Cigarette packs show a health warning that smoking can seriously harm and kill. Suggest **one** reason why people continue to smoke cigarettes. *(2 marks)*

People ignore the warning because they see other people who smoke and cannot see any obvious harm, so they think it won't harm them.

There are other possible answers than this, but your answer should clearly explain why people ignore obvious danger.

Worked example **target B-A***

Overuse of alcohol is the greatest cause of liver disease. A government report on alcohol and health in 2010 used the graph to argue for an increase in the cost of alcohol to reduce consumption. Use the graph to evaluate this argument. *(3 marks)*

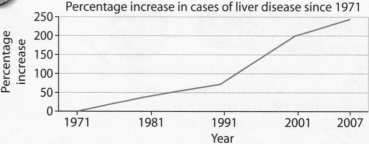

Percentage increase in cases of liver disease since 1971

The graph shows a 250% change in liver disease cases over 36 years. This suggests people are drinking more alcohol, but not why they drink more. If it is because alcohol is cheaper now, then increasing its cost might reduce drinking. If drinking patterns have changed for other reasons, then increasing cost may have no effect.

Now try this

 target D-C

1 (a) Alcohol is a recreational drug. Give **one** reason why it is classified as a legal drug. *(1 mark)*

(b) Give **one** reason why alcohol can be a problem. *(1 mark)*

 target B-A*

2 Some European countries are considering making tobacco smoking illegal, to reduce the cost of treating thousands of people each year who have smoking-related diseases. Evaluate this idea. *(3 marks)*

Drugs in sport

Some types of drugs can affect sporting performance. Examples include:

STIMULANTS that increase the rate of body functions, such as heart rate

A faster heart rate delivers oxygen and sugars to muscles more quickly, so they can release more energy more quickly.

ANABOLIC STEROIDS stimulate muscle growth.

Bigger muscles can help move bigger weights and generate more power, e.g. in weightlifting.

Drug bans in sport

Use of drugs to improve performance in sports competition is considered UNETHICAL.

Reasons for this include:

- side effects of the drugs can harm athletes
- the drugs may give an unfair advantage over athletes who don't use them.

Some drugs that affect performance are legal and some can be prescribed, but all are banned by sports competition regulations.

Ethical questions are about what people think is right or wrong. In a question about ethical issues, you may need to explain why different people have different ideas about what is right or wrong.

Worked example

target D-C

AQA SKILL
Interpret
Page 95

The graph shows the effect of injecting different doses of testosterone (a steroid) on 60 men who were weight training for 20 weeks. Use information from the graph to explain why testosterone is a banned drug in sports. *(2 marks)*

The graph shows that higher doses of testosterone increase the amount of weight that the leg muscle can lift. This could give an athlete an unfair advantage over athletes who haven't used testosterone.

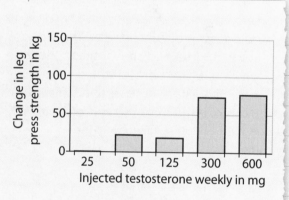

Now try this

target D-C

target B-A*

1 Explain why stimulants can improve sporting performance. *(2 marks)*

2 Erythropoietin (EPO) is a natural hormone that stimulates red blood cell production when oxygen levels are low, such as at high altitude. Red blood cells deliver oxygen to muscle cells and so can affect sporting performance. Artificial EPO was widely used by athletes until 2000 when a test was developed that could distinguish it from natural EPO. Athletes continue to train at high altitudes before major competitions. Evaluate the decision to make artificial EPO illegal for competing athletes to use. *(3 marks)*

Biology six mark question 2

There will be one 6 mark question on your exam paper which will be marked for *quality of written communication* as well as scientific knowledge. This means that you need to apply your scientific knowledge, present your answer in a logical and organised way, and make sure that your spelling, grammar and punctuation are as good as you can make them.

Worked example

Describe how new medical drugs are tested and explain why each stage of testing is needed.
(6 marks)

New drugs are first tested in the laboratory. They are tested on cells, tissues and live animals. This is to make sure they work properly.

If the first tests are successful, the drug is then tested in very low doses on healthy people to make sure the drug doesn't harm them.

Then the drug is tested on patients who have the disease that the drug is being developed to treat. The dose is increased to find the optimum dose, which has the best effect on the patient's illness. These tests may be double-blind trials using a placebo to make sure they are fair tests.

Organising your answer

Make sure you write your answer in a LOGICAL ORDER. There is often more than one way to organise an answer, and it does not matter which way you choose, as long as it is clear.

This part of the answer could have been made much clearer by explaining what 'properly' means. For example, it could have said that the drug is tested to make sure it has the effect that doctors intend, and to make sure that the drug is not toxic (poisonous).

A better answer would have explained that a double-blind trial means the doctor and patient don't know whether the patient has had the new drug or a placebo, which looks just like the drug but contains no drug. A double-blind trial means that the patient can't be affected by how they feel about getting either the drug or placebo, and that doctors are more objective about analysing the results.

Now try this

1 Describe how the process of IVF is carried out and use the graph to explain why IVF clinics are now advised to only transfer one embryo into the mother's womb. *(6 marks)*

Still birth means that the baby dies before it is born.

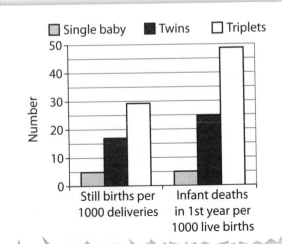

22

Competition

Organisms need a supply of materials from their surroundings, and sometimes from other living organisms, so that they can survive and reproduce. This means there is COMPETITION between organisms for materials that are in limited supply.

Competition between plants

competition for light and space

competition for water and nutrients

Competition between animals

Animals may compete with each other for:

- food
- mates for reproduction
- TERRITORY (space for feeding, reproduction and rearing young).

You will be expected to know the factors that organisms are competing for in an example.

Worked example target D-C

In spring, a male robin will sing loudly. Explain the role of singing in robins in terms of competition. *(2 marks)*

A male robin competes with other male robins for mates (females) and for territory. Singing loudly warns other males to keep out of the robin's territory and attracts females who choose to mate with him.

EXAM ALERT!

Make sure that you know what the command words mean. **Explain** means give a reason why. **Suggest** means you need to apply your knowledge to a new situation. **Describe** means say what is happening.

Students have struggled with questions like this in recent exams – **be prepared!**

Deterring predators

Many organisms are the food of other organisms. Some animals and plants have special features that deter PREDATORS.

Some animals advertise that they are poisonous with very bright colours.

Some animals use colours to make them look more frightening, like these big 'eyes'.

Some plants have big thorns.

Other plants are poisonous.

Now try this

 1 A farmer plants the seeds of his crop plants so that they are well separated from each other.
Explain why. *(2 marks)*

 2 Milk snakes are non-poisonous snakes. Coral snakes are highly poisonous. Milk snakes that live in the same area as coral snakes have a bold red and black pattern that is similar to that of the coral snakes. Suggest the advantage to the milk snake of this patterning. *(2 marks)*

Adaptations

All organisms (including microorganisms) have ADAPTATIONS that help them survive the conditions of the environment in which they normally live.

Animals in the Arctic

Small ears (reduced surface area) – less heat loss to air.

White colour for CAMOUFLAGE against snow.

Large feet for better grip on ice and stop bear sinking into snow.

Thick fur and fat below skin INSULATE (reduce rate of heat loss).

Animals in dry environments

Camels:
- have a hump of fat that is a food store; fat also releases water as a result of respiration
- can drink large quantities of water at a time
- have a thick coat at the top of the body that insulates against heat from the Sun.

Dry environments can also be very hot. African elephants have:
- large ears to transfer body heat to air quickly.

Worked example

target **D-C**

Label the diagram to describe how the listed adaptations help a cactus to survive in a dry environment.

(3 marks)

The cactus has no leaves which reduces water loss to air, but its green body still allows photosynthesis to take place.

Thick fleshy body stores water inside for times of drought.

Large root system collects as much water as possible from underground.

Extreme environments

Extreme environments have extreme conditions, for example:
- high levels of salt, e.g. a saltmarsh
- high temperatures, e.g. volcanic hot springs
- high pressures, e.g. the deep ocean.

Extremophiles

EXTREMOPHILES are organisms that have adaptations so they can TOLERATE very extreme environments. Many extremophiles are microorganisms.

Now try this

target **D-C**

1 The Arctic hare lives in areas where there is deep snow for many months, but the snow melts in summer. The fur of the hare is white in winter, but is replaced by brown fur in the summer. Explain why. *(2 marks)*

target **B-A***

2 Many deep ocean organisms can produce unique, flashing patterns of light. Suggest why they do this. *(2 marks)*

Indicators

Factors in the environment affect living organisms and their DISTRIBUTION (how widely spread they are). Changes in these factors may change their distribution.

environmental factors

living factors, e.g.
• prey
• competitor
• predator

non-living factors, e.g.
• light
• average temperature
• average rainfall
• oxygen levels in water
• pollution

Changes in these factors can affect organisms. For example, if there is a change in average temperature or rainfall this may change the distribution of organisms in an area.

Oxygen levels are high in unpolluted water and low in polluted water.

Changes in non-living factors can be measured using equipment, e.g. oxygen meter, thermometer, rainfall gauge.

Pollution indicators

Some species can be used as INDICATORS of pollution in air or in water.

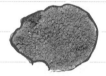

This LICHEN species indicates lots of air pollution, e.g. high sulfur dioxide concentration in air.

This lichen species indicates no air pollution, e.g. no sulfur dioxide in air.

Bloodworms indicate highly polluted water.

Mayfly larvae indicate unpolluted water.

Worked example

 D-C

AQA SKILL Analyse Page 95

The table shows the results of lichen surveys at the same place in two different years. Use the data in the table to suggest what happened to air quality between 1975 and 2010.   *(3 marks)*

Survey year	Number of lichens found on 5 trees	
	Lichens tolerant of high pollution	Lichens intolerant of high pollution
1975	35	1
2010	22	24

The numbers of lichens intolerant of high pollution increased by 230% between 1975 and 2010, and the numbers of lichens tolerant of high pollution decreased by 37% during this time. These results suggest that the level of air pollution in this place decreased between 1975 and 2010.

Where a question says 'use the table', make sure you refer to the information in the table in your answer.

Now try this

 D-C

1 Describe how invertebrate animals are used as water pollution indicators.   *(2 marks)*

2 Name **one** non-living factor that could be measured to indicate water pollution, and describe how its level changes with pollution.   *(2 marks)*

 B-A*

3 Compare the use of living and non-living methods of monitoring pollution.   *(3 marks)*

Energy and biomass

The source of energy for most food chains is light energy from the Sun.

light energy from Sun → small amount of energy captured by green plants and algae → **photosynthesis** *(light energy transferred to chemical energy)* → chemical energy stored in substances in cells and tissues

mass of living material = BIOMASS

Pyramids of biomass

The biomass of organisms at different levels of a food chain can be shown in a PYRAMID OF BIOMASS.

Worked example

 target **D–C**

AQA SKILL **Interpret** Page 95

The table shows the biomass of organisms in the food chain:

lettuce → caterpillar → thrush

Use the data in the table to draw a pyramid of biomass. *(3 marks)*

Organism	Biomass in g/m²
lettuce	120
caterpillar	60
thrush	12

Draw your diagram on graph paper. Make sure the width of each bar is drawn to scale.

thrush 12 g/m²
caterpillars 60 g/m²
lettuces 120 g/m²

The bars are always in the same order: start at the bottom of the food chain, working along the food chain as you move to the top of the pyramid.

Worked example

 target **B–A***

Explain the shape of a pyramid of biomass. *(3 marks)*

In animals, some materials and energy are lost to the environment in the waste they produce, including the carbon dioxide from respiration.

Also, not all the organisms in one level are eaten by the higher level. So there is less biomass in each level as you move up the pyramid.

Remember that energy and biomass are not the same thing. Biomass is broken down in respiration, releasing carbon dioxide, and some of the energy released from respiration is heat energy.

Now try this

 target **D–C**

1 A rabbit eats 500 g of grass. Explain why the rabbit will not increase in biomass by 500 g. *(2 marks)*

 target **B–A***

2 Explain a strength and a weakness of using a pyramid of biomass as a model of what is happening in a habitat. *(4 marks)*

26

Decay

There is a constant cycling of materials between living organisms and the environment. In a STABLE (unchanging) community, the amount of materials removed from the environment is balanced by the amount returned by decay.

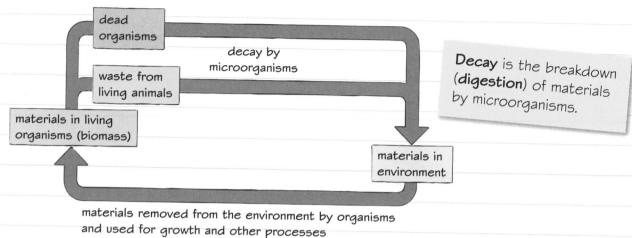

dead organisms

waste from living animals

decay by microorganisms

materials in living organisms (biomass)

materials in environment

materials removed from the environment by organisms and used for growth and other processes

Decay is the breakdown (**digestion**) of materials by microorganisms.

Recycling kitchen and garden waste

Garden and kitchen waste can be used to make COMPOST, either in the garden or by council schemes. Conditions in the compost should be controlled to encourage the growth of decay microorganisms, which grow and digest faster in conditions that are:

• moist   • warm   • aerobic (oxygen present).

 Worked example **D-C**

 AQA SKILL Analyse Page 95

The graph shows the amount of household waste collected by a council for some of the years between 2000 and 2012. It also shows how the waste was disposed of. The council's aim was to reduce the amount of waste sent to landfill.

Use the graph to help you explain how composting has helped the council to achieve its aim.

(2 marks)

The proportion of the waste sent for composting has increased each year. This means that less waste is left to go to landfill.

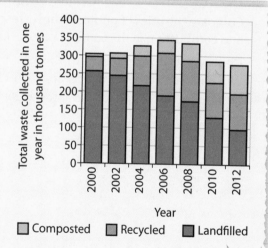

Total waste collected in one year in thousand tonnes

Year

☐ Composted   ☐ Recycled   ☐ Landfilled

 Now try this

 D-C

1 Why do compost heaps produce compost fastest in sunny places? *(1 mark)*

2 If a gardener grows crops in the same piece of ground each year, without adding compost, the yield (amount of food) he gets from the crops gets less and less. Explain why. *(2 marks)*

B-A*

3 Many local councils collect food and garden waste separately from other household refuse, and use it to make compost. Explain the value of these schemes. *(2 marks)*

Carbon cycling

The constant cycling of carbon between the air and living organisms is called the CARBON CYCLE.

Combustion is burning.

Detritus feeders are animals that eat dead and decaying material, e.g. earthworms.

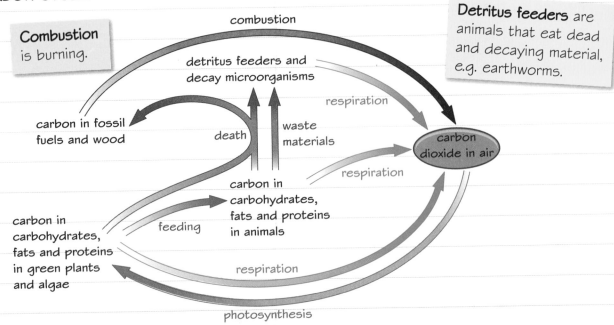

By the time the microorganisms and detritus feeders have broken down the waste products and dead bodies of organisms, and cycled the materials as plant nutrients, all the energy originally absorbed by green plants and algae has been transferred.

Worked example   target D–C

Compare the roles of photosynthesis, respiration and combustion in the carbon cycle.   *(3 marks)*

Photosynthesis is the process in which carbon dioxide is removed from the air and converted into carbon compounds in green plants and algae.

Respiration is the process that releases carbon dioxide from living organisms back into the air.

Combustion is the process that releases carbon dioxide from dead organisms back into the air.

EXAM ALERT!

Remember that plants photosynthesise only when it is light, but they also respire 24 hours a day. Microbes respire too.

Students have struggled with questions like this in recent exams – **be prepared!**

Now try this

target **D–C**

1 Describe how carbon in animal waste is returned to the air.   *(2 marks)*

target **B–A***

2 Combustion of fossil fuels is thought to be causing an increase in carbon dioxide concentration in the Earth's atmosphere.
   (a) Identify the source of carbon in fossil fuels.   *(1 mark)*
   (b) Explain fully why combustion may be causing an increase in carbon dioxide concentration in the air.   *(2 marks)*

Genes

Genes, chromosomes and nucleus

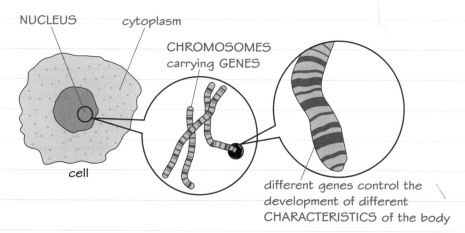

NUCLEUS   cytoplasm

CHROMOSOMES carrying GENES

cell

different genes control the development of different CHARACTERISTICS of the body

Worked example   target B-A*

Explain as fully as you can, in terms of genes, why individual plants and animals are similar but not identical to their parents.   *(3 marks)*

> Genes work at the molecular level to produce the characteristics that we see.

A plant or animal inherits half its genes from its father and half from its mother during fertilisation, when sex cells (gametes) from each parent fuse. Each of their genes controls the development of a particular characteristic. So the offspring have some characteristics from one parent and some from the other.

Causes of differences

differences between individuals of the same kind

Most variations are caused by a combination of genes and environment.

caused by differences in genes they have inherited, e.g. eye colour (GENETIC CAUSES)

combination of both causes, e.g. weight, skin colour

caused by differences in conditions in which they developed, e.g. riding a bike, scars (ENVIRONMENTAL CAUSES)

Now try this

target D-C

1 Name **two** different causes of differences in characteristics in dogs.   *(2 marks)*

target B-A*

2 Identical twins are produced when a fertilised egg cell divides in two on its first division. Use this information to explain why identical twins share many, but not all, of their characteristics.   *(2 marks)*

Reproduction

REPRODUCTION is the production of new individuals. There are two forms of reproduction.

1 Sexual reproduction

gamete from mother fuses (joins) with gamete from father

↓

mixes genetic information from each parent

↓

offspring have different combinations of genes, so show variety in characteristics

2 Asexual reproduction

no fusion of gametes – only one parent

↓

no mixing of genetic information

↓

all offspring have same genes as parent and each other

Worked example target D-C

There are two forms of reproduction. Which form of reproduction produces individuals that are genetically different from each other? Give a reason for your answer. *(2 marks)*

Sexual reproduction, because there is mixing of genes from the two parents during fertilisation.

Clones are individuals with identical genes.

Taking plant cuttings

New plants can be produced quickly and cheaply, by taking cuttings from an older plant.

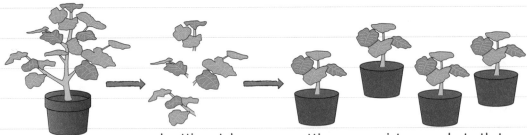

several cuttings taken from parent plant

cuttings grown into new plants that are genetically identical (clones)

Now try this

target D-C

1 A rose grower is growing many varieties of roses in a field. Each year she takes seed from some of the roses, plants them and looks after them until they are fully grown.
   (a) Will the rose plants grown from seed have the same shape and colour of flower as the other rose plants in the field? Explain your answer. *(2 marks)*
   (b) One year she finds a plant that produces beautiful flowers. Explain how she should produce these new plants. *(2 marks)*

Cloning

There are several modern techniques for cloning organisms.

 Tissue culture

Plant tissue culture is like taking plant cuttings, but with very small pieces that contain just a few plant cells. The new plants are all clones because their cells contain the same genes as the parent plant.

 Embryo transplants

Splitting embryos makes it possible to produce small numbers of clone animals at the same time. This is most often done with high-quality farm animals.

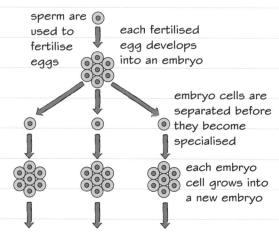

sperm are used to fertilise eggs

each fertilised egg develops into an embryo

embryo cells are separated before they become specialised

each embryo cell grows into a new embryo

- all the offspring are genetically identical (clones)
- each embryo is placed in the womb (uterus) of a different SURROGATE MOTHER to develop until ready for birth

 Adult cell cloning

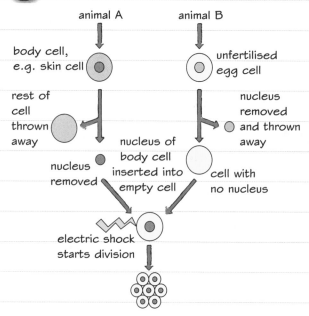

animal A   animal B

body cell, e.g. skin cell

unfertilised egg cell

rest of cell thrown away

nucleus removed and thrown away

nucleus removed

nucleus of body cell inserted into empty cell

cell with no nucleus

electric shock starts division

- cell divides to form embryo
- embryo inserted into womb of adult female C to develop until birth

---

 Worked example **D-C**

In the diagram shown above right, will the animal that is born be a clone of animal A, animal B or the adult female C in whose womb it developed? Explain your answer.   *(2 marks)*

Animal A. It has the same genes because the nucleus in the egg was taken from the cell of animal A.

 Now try this

 D-C

1 Describe **one** way in which plant tissue culture and taking plant cuttings are:
   (a) similar, **(b)** different.   *(2 marks)*

2 Describe the role of each of the three adult animals used in the development of a cloned offspring by adult cell cloning.   *(3 marks)*

B-A*

3 It costs more to use embryo transplanting to produce calves than to mate cows with a bull. Suggest why a farmer would use embryo transplanting to produce clones that develop inside surrogate mother cows, rather than get the bull to mate with each of the cows instead.   *(3 marks)*

Genetic engineering

Genetic engineering produces GENETICALLY MODIFIED (GM) organisms.

GENETIC ENGINEERING is the transfer of a gene from one organism to a different organism so that the desired characteristic is produced in that organism.

The jellyfish has a gene that produces a chemical that glows in blue light. A mouse doesn't normally have this chemical. This glowing mouse has been genetically modified.

Genes can be transferred from any kind of organism to any other kind of organism, e.g. bacteria, humans, other animals, plants.

| the gene for a characteristic is 'cut out' of a chromosome using enzymes | → | the gene is inserted into a chromosome inside the nucleus of a cell in a different organism | → | the cell of this organism now produces the characteristic from the gene |

Worked example   B-A*

AQA SKILL
Explain
Page 95

Human insulin is a hormone used by many patients with diabetes. Bacteria have been genetically modified to carry the human insulin gene. The bacteria are then grown in large quantities to produce the human insulin.

Explain why bacteria had to be modified to make human insulin and describe how the GM bacteria were produced.   *(3 marks)*

Bacteria don't normally produce human insulin, so they had to be modified to make it. The gene for making insulin was cut out of a human chromosome. The gene was then inserted into the bacterial chromosome, so that the bacterium made the insulin.

GM crops

GM CROP plants have been genetically modified to give them new characteristics, such as:

- resistance to attack by insects
- resistance to HERBICIDES, so that fields can be sprayed to kill weeds, but not the crop.

These characteristics can help the crop grow better and produce more food (an increased YIELD).

EXAM ALERT!

You are expected to know the different steps in the process of genetic modification.

Students have struggled with questions like this in recent exams – **be prepared!**

Now try this

target **D-C**

1   Describe how a GM crop with herbicide resistance could be developed.   *(2 marks)*

target **B-A***

2   The production of 'glow mice' has a serious purpose, because the glow gene is joined to a gene that causes a human genetic disease. These mice are then used for testing new treatments for the disease.
   (a) Suggest **one** advantage of using a glow gene joined to the disease gene.   *(1 mark)*
   (b) Describe how lots of mice with the human disease gene could be produced.   *(2 marks)*

Issues with new science

New scientific developments cause new issues that we need to think about. To make good judgements about these developments we need good information.

economic issues: related to money	social issues: related to people and society	ethical issues: related to right and wrong, or fairness
↓	↓	↓
judgement: is it worth the cost?	judgement: are the benefits for people/society greater than the problems?	judgement: is it right/wrong/fair to do this?

If you are asked a question on genetic engineering or adult cell cloning, the example used may not be something you have studied in class. Don't panic. All the information you need to answer the question will be on the paper. If you are asked to make an informed judgement you could think about the different kinds of issues shown on the flow charts.

Concerns about GM crops

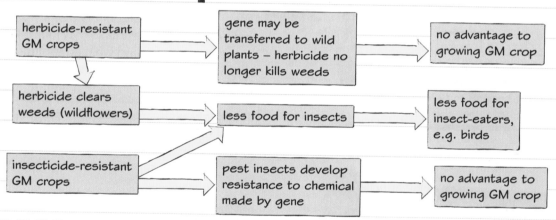

herbicide-resistant GM crops	→	gene may be transferred to wild plants – herbicide no longer kills weeds	→	no advantage to growing GM crop
herbicide clears weeds (wildflowers)	→	less food for insects	→	less food for insect-eaters, e.g. birds
insecticide-resistant GM crops	→	pest insects develop resistance to chemical made by gene	→	no advantage to growing GM crop

Worked example D-C

Foods made from GM crops have been sold in the US for over 10 years. Scientific studies show no evidence that GM foods harm health. Suggest why some people are still concerned about the possible effects on health of eating GM foods. *(2 marks)*

There is a possibility that eating these foods over a long time may harm health. These effects would not show yet because GM foods have only been available for a few years.

Now try this

D-C

1 If the gene for herbicide resistance is transferred to weed plants, there will be no advantage in growing the GM crop. Describe the economic impact of this. *(2 marks)*

B-A*

2 Using GM crops that are herbicide-resistant makes it easier for a farmer to keep a crop free of weeds. Explain why some people are concerned that this will lead to the extinction of many wild animal species that live on farmland. *(4 marks)*

Evolution

EVOLUTION means change over time.

Classification

Organisms are CLASSIFIED as plant, animal or microorganism using the similarities and differences in their characteristics.

Classification can show how organisms are related by evolution or by how they live (ecological relationship).

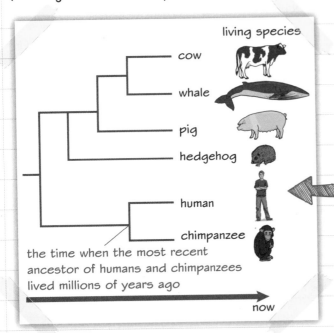

living species

- cow
- whale
- pig
- hedgehog
- human
- chimpanzee

the time when the most recent ancestor of humans and chimpanzees lived millions of years ago

now

The evidence for evolution suggests that all living species have evolved from simple life forms that first developed over 3 billion years ago.

Models of evolutionary relationships

EVOLUTIONARY TREES are models of the relationships between organisms.

The tree on the left shows that cows and whales have more similar characteristics than cows and humans, for example. This suggests that cows and whales evolved from the same ancestor more recently than cows and humans.

EXAM ALERT!

Different kinds of evolutionary tree are based on different evidence and so may look different. You need to understand what an evolutionary tree shows, so that you can interpret it, no matter what it looks like.

Students have struggled with this topic in recent exams – **be prepared!**

Worked example   target D-C

AQA SKILL Suggest Page 95

The diagram shows the forelimbs of two mammals. Suggest how this evidence supports the idea that these mammals evolved from the same ancestor. *(2 marks)*

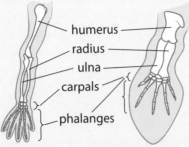

- humerus
- radius
- ulna
- carpals
- phalanges

human arm – for holding and working with objects

whale flipper – for swimming underwater

The bones in each limb are arranged in the same way. This suggests they evolved from the same ancestor with this bone arrangement.

Now try this

target D-C

1 Look at the evolutionary tree above.
   (a) What does the tree show about how similar humans, cows and pigs are? *(2 marks)*
   (b) Which of these pairs of animals shares the most recent common ancestor: cow/pig, cow/hedgehog? Explain your answer. *(2 marks)*

target B-A*

2 In 1758 Linnaeus classified earthworms, sea slugs, hagfish and other animals as Vermes, because they moved slowly, were soft, could replace parts that broke off and lived in moist places. Today these animals are classified in different groups using many more features. Suggest why the classification has changed. *(3 marks)*

Theories of evolution

There have been many different theories about how evolution happens. This page shows two theories that you need to know about.

 Darwin's theory

Darwin's theory states that evolution happens through NATURAL SELECTION.

Natural selection works on inherited characteristics (caused by GENES).

Relatively rapid evolution can occur if:

- a new form of a gene occurs due to MUTATION
- there is a large change in the environment.

| individuals in a species vary in their characteristics – some of these variations are caused by genes | → | some variations are better suited to the environment than others | → | individuals with the better-suited variations are more likely to breed successfully and pass on genes for the better-suited variations | → | the next generation will contain more individuals with the genes for better-suited variations |

 Lamarck's theory

Long before Darwin, Lamarck thought evolution was due to the inheritance of characteristics changed by the environment.

Darwin's theory explains evolution better than Lamarck's theory in most cases. You need to learn the differences between Darwin's and Lamarck's theories.

| environmental factor, e.g. weight training | → | causes an **acquired characteristic** e.g. increased muscle size | → | acquired characteristic inherited by offspring, e.g. children have big muscles |

 Worked example **D-C**

AQA SKILL
Explain
Page 95

Explain **one** reason why Darwin's theory was only gradually accepted. *(2 marks)*

At the time that Darwin published his theory, many people believed God had created all the species. So they didn't believe evolution could happen just by natural selection.

Other possible answers are:

- Darwin's theory was published about 50 years before the theory about genes, so he couldn't explain how variations were passed on to offspring.
- Evolution happens slowly, so it can take a long time to collect evidence that evolution has happened.

 Now try this

target **D-C**

1 Jack and his daughter are both champion swimmers.
   (a) Suggest how Lamarck would have explained the evolution of being a champion swimmer. *(2 marks)*
   (b) Suggest how Darwin would have explained this. *(2 marks)*

 target **B-A***

2 Herbicide-resistant weeds are becoming a major problem in areas where GM herbicide-resistant crops are grown. Explain why these weeds have only been discovered in these areas and why they have appeared so rapidly. *(2 marks)*

35

Biology six mark question 3

There will be one 6 mark question on your exam paper which will be marked for *quality of written communication as well as scientific knowledge*. This means that you need to apply your scientific knowledge, present your answer in a logical and organised way, and make sure that your spelling, grammar and punctuation are as good as you can make them.

Worked example

In parts of Africa the maize crops are badly damaged each year by pests. This leaves little food for the farmers and their families to eat.

A new variety of genetically modified maize has been produced using a gene from a bacterium, and this new variety is resistant to the pests. The seed for the GM maize is more expensive than seed for normal maize, but it guarantees a good yield.

Describe how the GM maize was produced, and explain some of the arguments for and against GM crops. *(6 marks)*

The gene for resistance to the pests was taken out of the bacterium. The gene was inserted into the maize plants to make them resistant to the pests.

Arguments for growing GM maize are that the maize plants will produce a greater yield, which means that the farmers and their families will have more food to eat.

Arguments against growing the GM maize are that it is more expensive than growing normal maize, and some people think that GM food could be harmful to health.

EXAM ALERT!

Remember that these questions take into account how good your punctuation is. Make sure that you start each sentence with a capital letter and end with a full stop.

Students have struggled with questions like this in recent exams – **be prepared!**

This answer gives two different points about how a GM plant could be made, but it really should explain that the gene was put into a very early stage of a maize plant, so that all the cells of the adult plant contained the gene.

This question doesn't tell you to use only what's written to help you write your answer. So you could give other arguments for and against that you remember from your course, such as people are worried that GM crops harm the wildflowers and insects other than the pests, or that pests are evolving so they can grow on these resistant crops.

Now try this

1 Describe the roles of different organisms in the carbon cycle.

decomposers   primary consumer   secondary consumer   producer   *(6 marks)*

Atoms and elements

Elements

All substances are made of ATOMS. An individual atom is too small for you to see, so everything around you contains very many atoms.

An ELEMENT is a substance that is made of only one sort of atom. Oxygen is an element because it only contains oxygen atoms.

Chemical symbols

There are about 100 different elements. Atoms of each element are given a chemical symbol.

Every symbol starts with a capital letter, usually followed by a lower case letter. For example, N represents a nitrogen atom, but Na represents a sodium atom.

The elements are shown in the PERIODIC TABLE.

Periodic table

group numbers — 1   2 ←   → 3   4   5   6   7   0

1	2											3	4	5	6	7	0
							1 H Hydrogen 1		Non-metals on the right								**4 He** Helium 2
7 Li Lithium 3	**9 Be** Beryllium 4	Each group (a vertical column) contains elements with similar properties										**11 B** Boron 5	**12 C** Carbon 6	**14 N** Nitrogen 7	**16 O** Oxygen 8	**19 F** Fluorine 9	**20 Ne** Neon 10
23 Na Sodium 11	**24 Mg** Magnesium 12						Metals on the left					**27 Al** Aluminium 13	**28 Si** Silicon 14	**31 P** Phosphorus 15	**32 S** Sulfur 16	**35.5 Cl** Chlorine 17	**40 Ar** Argon 18
39 K Potassium 19	**40 Ca** Calcium 20	**45 Sc** Scandium 21	**48 Ti** Titanium 22	**51 V** Vanadium 23	**52 Cr** Chromium 24	**55 Mn** Manganese 25	**56 Fe** Iron 26	**59 Co** Cobalt 27	**59 Ni** Nickel 28	**64 Cu** Copper 29	**65 Zn** Zinc 30	**70 Ga** Gallium 31	**73 Ge** Germanium 32	**75 As** Arsenic 33	**79 Se** Selenium 34	**80 Br** Bromine 35	**84 Kr** Krypton 36
85 Rb Rubidium 37	**88 Sr** Strontium 38	**89 Y** Yttrium 39	**91 Zr** Zirconium 40	**93 Nb** Niobium 41	**96 Mo** Molybdenum 42	**99 Tc** Technetium 43	**101 Ru** Ruthenium 44	**103 Rh** Rhodium 45	**106 Pd** Palladium 46	**108 Ag** Silver 47	**112 Cd** Cadmium 48	**115 In** Indium 49	**119 Sn** Tin 50	**122 Sb** Antimony 51	**128 Te** Tellurium 52	**127 I** Iodine 53	**131 Xe** Xenon 54
133 Cs Caesium 55	**137 Ba** Barium 56	**139 La** Lanthanum 57	**178 Hf** Hafnium 72	**181 Ta** Tantalum 73	**184 W** Tungsten 74	**186 Re** Rhenium 75	**190 Os** Osmium 76	**192 Ir** Iridium 77	**195 Pt** Platinum 78	**197 Au** Gold 79	**201 Hg** Mercury 80	**204 Tl** Thallium 81	**207 Pb** Lead 82	**209 Bi** Bismuth 83	**210 Po** Polonium 84	**211 At** Astatine 85	**222 Rn** Radon 86
223 Fr Francium 87	**226 Ra** Radium 88	**227 Ac** Actinium 89	**261 Rf** Rutherfordium 104	**262 Db** Dubnium 105	**266 Sg** Seaborgium 106	**264 Bh** Bohrium 107	**277 Hs** Hassium 108	**268 Mt** Meitnerium 109	**271 Ds** Darmstadtium 110	**272 Rg** Roentgenium 111	Elements with atomic numbers 112 – 118 have been reported but not fully authenticated						

Worked example   D-C

Name the particles found in an atom, and state their relative electrical charges. *(3 marks)*

Protons have a relative charge of +1, neutrons have a relative charge of 0, and electrons have a relative charge of −1.

Atoms have a small central **nucleus** made of positively charged **protons** and neutral **neutrons**.

There are negatively charged **electrons** around the nucleus.

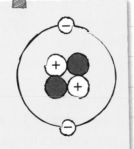

Now try this   D-C

1. Sodium and oxygen are both elements. Sodium is a metal.
   (a) State why sodium is called an element. *(1 mark)*
   (b) Why does oxygen have different properties than sodium? *(1 mark)*
2. Name the parts labelled A, B and C on the diagram of an atom. *(3 marks)*

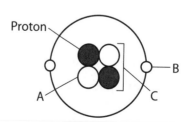

Proton

Particles in atoms

Protons, neutrons and electrons are called SUBATOMIC PARTICLES. You can work out how many of each type of subatomic particle an atom has from its atomic number and mass number.

Atomic number

The number of protons in an atom of an element is called its ATOMIC NUMBER.

The atoms of different elements have different numbers of protons – no two elements can have the same atomic number.

Number of electrons

Atoms have no overall charge. This is because the number of electrons in an atom is the same as the number of protons.

Mass number

The total number of protons and neutrons in an atom is called its MASS NUMBER.

Worked example **D-C**

A sodium atom has an atomic number of 11 and a mass number of 23. How many of each type of subatomic particle does it have?

(3 marks)

Number of protons = 11

Number of electrons = 11

Number of neutrons = 12

The number of protons is given by the atomic number, which is 11 for sodium.

The number of electrons in an atom is the same as the number of protons, which is 11 here.

The number of neutrons equals the mass number minus the atomic number.

So the number of neutrons = 23 – 11 = 12.

Diagrams of atoms

Atoms of elements can be drawn like the one in this diagram. This means that it is easy to work out how many subatomic particles there are.

In this particular diagram, the crosses represent electrons. They are arranged around the central nucleus.

If the atom has three electrons, it must also have three protons. This means that the black circles represent protons here.

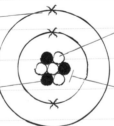

The white circles here represent neutrons.

The atomic number is 3 and the mass number is 7 (3 protons + 4 neutrons).

Now try this

 D-C

1   The table gives information about two atoms.

Atom	Mass number	Atomic number
X	40	20
Y	40	19

(a) Calculate the number of protons, neutrons and electrons in atom Y.   (3 marks)

(b) Explain whether the two atoms belong to the same element.   (2 marks)

Electronic structure

You should be able to represent the electronic structure of the first 20 elements.

Energy levels

The electrons in an atom occupy different ENERGY LEVELS around the nucleus. Each electron in an atom is at a particular energy level. Electrons occupy the lowest available energy levels.

Shells

Energy levels are also called SHELLS:
- The innermost shell is the lowest energy level.
- The outer shell is the highest occupied energy level.

Writing electronic structures

Different energy levels can contain different numbers of electrons. For the first 20 elements (hydrogen to calcium):

Energy level	Number of electrons
first	1 or 2
second	up to 8
third	up to 8
fourth	1 or 2

For example, a sodium atom has 11 electrons:
- 2 fit into the first energy level
- 8 fit into the second energy level
- 1 fits into the third energy level.

This electronic structure is written as 2,8,1 (the commas separate each energy level).

Use your periodic table to work out electronic structures. Count from hydrogen to the required element, for example sodium.

Electronic structures as diagrams

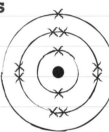

This is the electronic structure of sodium as a diagram.

Worked example D-C

(a) Write down the electronic structure of oxygen.

2,6

(b) Complete the diagram to show the electronic structure of oxygen (atomic number 8).   (2 marks)

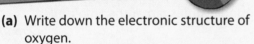

EXAM ALERT!

Make sure you can write and draw electronic structures correctly. The periodic table on the Data Sheet can help you:
- The number of energy levels (or circles) must be the same as the row the element is in.
- The total number of electrons must be the same as the element's atomic number.
- The last number must be the same as the element's group number.

Students have struggled with questions like this in recent exams – **be prepared!**

Now try this

1 Write down the electronic structures for atoms of the following elements:
   (a) carbon (atomic number 6)   *(1 mark)*
   (b) sulfur (atomic number 16)   *(1 mark)*
   (c) calcium (atomic number 20).   *(1 mark)*

2 Complete the diagram to show the electronic structure of aluminium (atomic number 13).   *(2 marks)*

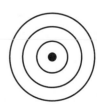

Electronic structure and groups

Atoms of the elements in a group in the periodic table have the same number of electrons in their highest energy level (outer shell). This gives the elements similar chemical properties.

Group 1 (alkali metals)

The elements in Group 1 include lithium, sodium and potassium. Their atoms all have just one electron in their highest occupied energy level (outer shell).

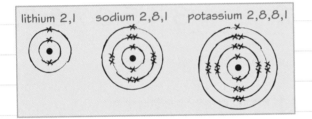

lithium 2,1   sodium 2,8,1   potassium 2,8,8,1

Group 0 (noble gases)

The elements in Group 0 include helium, neon and argon. The highest occupied energy levels (outer shells) of their atoms are full:
- Helium has two electrons in its highest occupied energy level.
- The others all have eight electrons in their highest occupied energy levels.

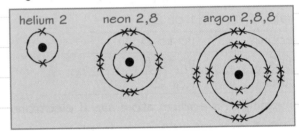

helium 2   neon 2,8   argon 2,8,8

Reactions with oxygen

These elements burn vigorously in oxygen to produce solid metal oxides. For example:

sodium + oxygen → sodium oxide

$4Na + O_2 \rightarrow 2Na_2O$

Reactions with water

These elements react vigorously with water to produce hydrogen gas, and metal hydroxides that dissolve to form alkaline solutions. For example:

sodium + water → sodium hydroxide + hydrogen

$2Na + 2H_2O \rightarrow 2NaOH + H_2$

Worked example B-A*

Gases such as hydrogen and oxygen exist as molecules in which two atoms are joined together. Explain in terms of electronic structure why the Group 0 elements exist as single atoms rather than molecules.
(2 marks)

The Group 0 elements are all unreactive, so their atoms do not join together. This is because their outer electron shells are full.

Now try this

1 In terms of electronic structure:
   (a) Why are noble gases unreactive? *(1 mark)*
   (b) Why do the elements in Group 1 have similar chemical properties? *(1 mark)*

2 How many electrons in there in the outer energy levels of the elements in Group 0? *(2 marks)*

3 Potassium reacts with water in a similar way to sodium. Write down a word equation and balanced symbol equation for the reaction. *(3 marks)*

The Data Sheet on page 94 may help when balancing equations.

Making compounds

When elements react with each other, their atoms join together to form COMPOUNDS.

Forming ions

Metals and non-metals react together to form compounds. The compounds are made of IONS.

An ion is a charged particle formed when an atom, or group of atoms, loses or gains electrons.

Forming molecules

Compounds formed from reactions between non-metals consist of MOLECULES.

In a molecule, electrons are shared between atoms. These shared electrons make COVALENT BONDS, which hold the atoms in the molecule together.

Giving and taking electrons

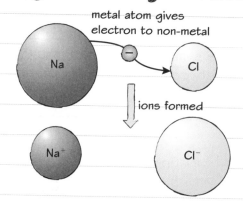

metal atom gives electron to non-metal

ions formed

Some examples of molecules

Carbon dioxide, CO_2 (a compound of carbon and oxygen)

Water, H_2O (a compound of hydrogen and oxygen)

Worked example

target B-A*

Potassium and bromine react together to form potassium bromide.

(a) Balance the symbol equation for the reaction. *(2 marks)*

$2K + Br_2 \rightarrow 2KBr$

(b) Explain how a potassium atom forms an ion. *(2 marks)*

A potassium atom loses one electron to form an ion with a charge of 1+.

Potassium is a metal and bromine is a non-metal. When they react together, potassium atoms give electrons to bromine atoms, forming ions.

Remember:
• metals lose electrons to form positive ions
• non-metals gain electrons to form negative ions.

The oppositely charged ions attract each other in compounds containing ions.

A bromine atom gains one electron to form an ion with a charge of 1−.

Now try this

target D-C

1 Explain how non-metals form negative ions. *(2 marks)*

2 Describe how a sodium ion forms. *(2 marks)*

3 (a) State the type of particles found in lithium iodide. *(1 mark)*
   (b) State the type of chemical bond found in molecules of methane, CH_4. *(1 mark)*

target B-A*

4 In terms of particles, describe how potassium chloride forms from its elements. *(3 marks)*

Chemical equations

We use word equations and balanced symbol equations to represent chemical reactions.

Word equations

In a chemical reaction:

- REACTANTS are the substances that react together
- PRODUCTS are the substances made.

In a WORD EQUATION, two or more reactants or products are separated by a + sign.

An example word equation

This word equation shows that iron oxide reacts with carbon to make iron and carbon monoxide:

> iron oxide + carbon → iron + carbon monoxide
> (reactants)   (products)

Take care to show all the reactants on the left of the arrow and all the products on the right.

Symbol equations

In a SYMBOL EQUATION, the chemical FORMULA of each reactant and product is shown, instead of its name.

In a correctly balanced symbol equation, there are the same number of atoms of each element in the reactants and in the products.

Here is the symbol equation for the reaction between iron oxide and carbon:

$$Fe_2O_3 + 3C \rightarrow 2Fe + 3CO$$

The symbol equation shows that each side has 2 Fe atoms, 3 O atoms and 3 C atoms.

Worked example   B-A*

Carbon reacts with oxygen to make carbon dioxide:
$C + O_2 \rightarrow CO_2$

12 g of carbon reacts with oxygen to form 44 g of carbon dioxide. Calculate the mass of oxygen needed to form 11 g of carbon dioxide.   (2 marks)

mass of oxygen for 44 g = 44 − 12 = 32 g
mass of oxygen for 11 g = 32 × 11/44 = 8 g

In chemical reactions, no atoms are lost and no new atoms are made.

This means that the total mass of the products is the same as the total mass of the reactants.

Now try this

D-C 1 Calcium carbonate breaks down when heated:
  calcium carbonate → calcium oxide + carbon dioxide
           50 g   ?   22 g
  Calculate the mass of calcium oxide made from 25 g of calcium carbonate.   (2 marks)

D-C 2 Explain, in terms of the number of atoms, why this is a correct symbol equation.   (2 marks)
           $H_2 + F_2 \rightarrow 2HF$

B-A* 3 Correctly balance these equations.
  (a) $CaCO_3 + \ldots HCl \rightarrow CaCl_2 + H_2O + CO_2$ (1 mark)
  (b) $\ldots Na + Cl_2 \rightarrow \ldots NaCl$ (1 mark)
  (c) $\ldots Al + \ldots O_2 \rightarrow \ldots Al_2O_3$ (1 mark)

Place a whole number in each of the spaces marked as ... Remember: an equation is balanced when there are equal numbers of atoms of each element on both sides of the equation.

Limestone

LIMESTONE is a type of rock. It is mostly calcium carbonate, $CaCO_3$.

Uses of limestone

As raw material for making cement, mortar and concrete.

As blocks and slabs for walls and pavements.

As AGGREGATE (small lumps) for the base of roads and railways.

Building materials

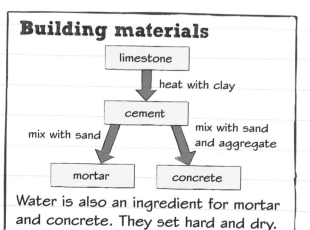

Water is also an ingredient for mortar and concrete. They set hard and dry.

Quarries

Limestone is taken from the ground in QUARRIES. Obtaining and using limestone has economic and social effects. For example:

✓ New jobs are created and limestone is a valuable material for building.

✗ Fewer tourists may visit an area that has a working quarry.

✗ Extra traffic.

✗ Noise, visual and atmospheric.

Limestone, cement and concrete are needed for buildings. Their positive effects must be considered against the negative effects of a quarry.

 Worked example — target **B-A***

 AQA SKILL **Evaluate** Page 95

The front of a building has two old limestone columns that need replacing. Evaluate the use of limestone or steel-reinforced concrete to replace them. *(4 marks)*

Limestone is an abundant natural material that is easy to shape. On the other hand, the raw materials for concrete are also abundant, and it can be moulded on site. I think limestone is a better choice because it is more attractive.

The answer describes two advantages of each material, then completes with a conclusion. It could have discussed disadvantages instead. For example, limestone is more easily damaged by acid rain than concrete.

Now try this

 target **D-C**

1 Name **two** substances that are mixed with cement to make concrete. *(2 marks)*

2 Describe **two** advantages and **two** disadvantages of quarrying limestone. *(4 marks)*

 target **B-A***

3 Cement is made by heating limestone and clay to 1500 °C using coal or natural gas.
 (a) Apart from the cost of raw materials, suggest **one** major expense in this process. *(1 mark)*
 (b) Name **one** polluting gas produced, and explain how it harms the environment. *(2 marks)*

Calcium carbonate chemistry

Thermal decomposition

Many metal carbonates break down when they are heated. The reaction is called THERMAL DECOMPOSITION.

metal carbonate	→	metal oxide	+	carbon dioxide

When calcium carbonate is heated, it decomposes to form calcium oxide and carbon dioxide:

$$CaCO_3 \rightarrow CaO + CO_2$$

Other carbonates

These carbonates decompose in a similar way to calcium carbonate when heated:
- magnesium carbonate
- zinc carbonate
- copper carbonate.

Not all carbonates of metals in Group I decompose when heated with a Bunsen flame.

Making an alkali

Calcium oxide reacts vigorously with drops of water to make a white solid called calcium hydroxide.

$$CaO + H_2O \rightarrow Ca(OH)_2$$

A lot of heat is given out in the reaction. Calcium hydroxide is an ALKALI.

Alkalis dissolve in water to make **alkaline** solutions. They **neutralise** acids.

Limewater

LIMEWATER is a solution of calcium hydroxide in water. It is used as a test for carbon dioxide. Carbon dioxide turns limewater cloudy (because tiny white particles of calcium carbonate form in the reaction):

calcium hydroxide + carbon dioxide	→	calcium carbonate + water

$$Ca(OH)_2 + CO_2 \rightarrow CaCO_3 + H_2O$$

Worked example

 target B-A*

A statue is made from limestone. It has been damaged by acid rain. Complete and balance this equation. *(2 marks)*

$$CaCO_3 + 2HCl \rightarrow CaCl_2 + CO_2 + H_2O$$

Excess acid in lakes and soil can be neutralised by adding powdered limestone, which is mainly calcium carbonate.

Other carbonates react with acids, including:
- sodium carbonate
- magnesium carbonate
- zinc carbonate
- copper carbonate.

Any acid reacts with a carbonate to produce a salt, carbon dioxide and water.

Now try this

 target D-C

1 (a) Explain why bubbles are seen when sodium carbonate is mixed with an acid. *(2 marks)*
  (b) Describe a laboratory test to identify these bubbles. *(2 marks)*

2 Write down a word equation for the reaction between calcium oxide and water. *(1 mark)*

 target B-A*

3 When heated, copper carbonate decomposes in a similar way to calcium carbonate.
  (a) Name the solid produced in the reaction. *(1 mark)*
  (b) Complete and balance the symbol equation for the reaction between copper carbonate and hydrochloric acid:
  $$CuCO_3 + \ldots HCl \rightarrow \ldots + \ldots$$ *(2 marks)*

Chemistry six mark question 1

There will be one 6 mark question on your exam paper which will be marked for *quality of written communication as well as scientific knowledge*. This means that you need to apply your scientific knowledge, present your answer in a logical and organised way, and make sure that your spelling, grammar and punctuation are as good as you can make them.

Worked example

AQA SKILL
Describe
Page 95

Limestone is important for building but to extract it from the Earth it must be quarried. Explosives and heavy machinery are used to get the limestone out of the ground and move it. Large lorries take the crushed limestone to the cement factory. The machinery and lorries use diesel fuel.

A quarry company wants to open a new limestone quarry in an attractive countryside area. Describe the positive and negative impacts of quarrying limestone there. *(6 marks)*

A limestone quarry will create new jobs in the area. This will improve the local economy and may attract other new businesses into the area. Limestone is a useful building material, and cement can be used to make mortar and concrete. These are valuable materials.

However, a limestone quarry will damage the environment. It will be noisy and look unpleasant. The habitats of local wildlife will be destroyed. The machinery and lorries will release smoke and carbon dioxide.

A balanced answer

The information given will help you with your answer, but you should not just repeat it.

It is important to give detailed descriptions of some advantages *and* disadvantages of the quarry.

Large lorries are used to transport the crushed limestone.

The answer could also have mentioned how the company might restore the land after the quarry closes, or improve local facilities such as roads.

The use of 'however' is a good way of moving to the disadvantages. Note that it gives specific examples of the damage to the environment.

Now try this

Limestone and brick are often used for constructing the outside walls of buildings. Use the information in the table, and what you know about limestone and brick, to explain the advantages and disadvantages of using limestone as a building material. *(6 marks)*

Building material	Cost in £ per m²	Energy needed to extract and process material in MJ/kg	Resistance to air pollution	Life span in years
Limestone	52	0.85	Medium	50 or more
Brick	39	3.00	High	50 or more

Extracting metals

Unreactive metals such as gold are found in the Earth's crust as the metal element itself. However, most metals are found as compounds. These need chemical reactions to extract them from their ores.

Ores

Rocks contain metals or their compounds. An ORE is a rock that contains enough of a metal to make its extraction economical.

Rocks may contain too little metal to make extraction worthwhile (if the cost of extracting the metal is greater than the value of the metal itself). Over time, metal prices may rise and these LOW-GRADE ORES may become useful.

Reduction

Iron is extracted from iron oxide in a blast furnace by reaction with carbon:

$$2Fe_2O_3 + 3C \rightarrow 4Fe + 3CO_2$$

This reaction works because carbon is more reactive than iron.

Other metals can be extracted like this if they are less reactive than carbon.

A reaction where oxygen is removed from a compound is called a REDUCTION reaction.

Reactivity and extraction

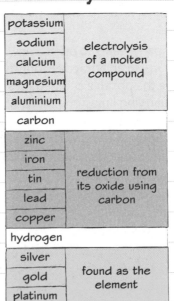

potassium	electrolysis of a molten compound
sodium	
calcium	
magnesium	
aluminium	
carbon	
zinc	reduction from its oxide using carbon
iron	
tin	
lead	
copper	
hydrogen	
silver	found as the element
gold	
platinum	

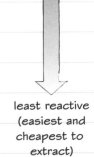

most reactive (most difficult and expensive to extract)

least reactive (easiest and cheapest to extract)

Worked example   target D–C

Aluminium must be extracted by electrolysis of molten aluminium oxide.

(a) State why reduction with carbon cannot be used. *(1 mark)*

Carbon is less reactive than aluminium so it cannot remove oxygen from aluminium oxide.

(b) Suggest **two** reasons why aluminium extraction is expensive. *(2 marks)*

Large amounts of energy are needed to melt aluminium oxide, and large amounts of electrical energy are needed for electrolysis.

Titanium cannot be extracted by reduction with carbon, either. As with aluminium, a lot of energy and many stages are needed.

EXAM ALERT!

The reactivity series of metals is on the Data Sheet given in the exam, so you do not need to memorise it.

Students have struggled with exam questions similar to this – **be prepared!**

Now try this

target D–C

1 Suggest which method would be used to extract potassium. *(2 marks)*

2 Suggest why iron is extracted using carbon, rather than by the electrolysis of iron oxide. *(2 marks)*

target B–A*

3 Zinc can be extracted from zinc oxide, ZnO, by reaction with carbon in a blast furnace. Write down a balanced equation for the reaction. *(2 marks)*

Extracting copper

High-grade copper ores contain a high proportion of copper compounds.

Smelting

Copper is extracted from these ores by SMELTING. This involves heating copper ores in a furnace. For example, copper sulfide is heated in air to produce copper:

$$CuS + O_2 \rightarrow Cu + SO_2$$

Traditional mining and extraction methods have major environmental impacts. High-grade ores are running out, so other extraction methods are being researched.

Electrolysis

Copper is purified by ELECTROLYSIS.

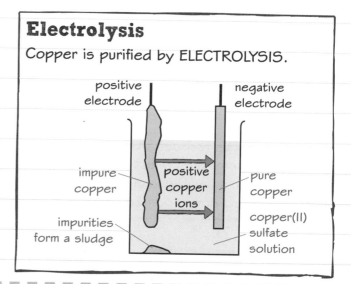

positive electrode / negative electrode
impure copper / positive copper ions / pure copper
impurities form a sludge / copper(II) sulfate solution

Worked example   **D-C**

Scrap iron can be used to produce copper from solutions of copper salts.

(a) Explain why this happens.   *(2 marks)*

Iron is more reactive than copper. This means it can displace copper from solutions of copper compounds.

(b) Suggest why the use of scrap iron is an economical way to produce copper.
   (1 mark)

Scrap iron is cheap.

In a **displacement** reaction, a more reactive metal can displace a less reactive metal from solutions of its compounds. For example:

Fe + CuSO₄ → FeSO₄ + Cu

Scrap iron is more abundant than copper.

Phytomining

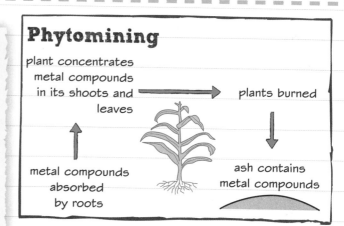

plant concentrates metal compounds in its shoots and leaves → plants burned
metal compounds absorbed by roots
ash contains metal compounds

Bioleaching

Copper can also be extracted by BIOLEACHING. Certain bacteria absorb metal compounds to produce a solution called LEACHATE. This has a high concentration of metal compounds. Scrap iron may be used to displace copper metal from these solutions.

Now try this

D-C

1. Explain why supplies of copper may be limited in the future.   *(2 marks)*

2. Copper can be extracted from copper(II) sulfate solution using electrolysis. Explain where the copper would collect.   *(2 marks)*

B-A*

3. **(a)** During smelting, copper oxide is produced from copper sulfide. Balance this symbol equation: $Cu_2S + \ldots\ldots O_2 \rightarrow \ldots\ldots CuO + SO_2$ *(1 mark)*
   (b) Explain why iron can be used to extract copper from a solution.   *(2 marks)*
   (c) Give **two** advantages of phytomining over traditional extraction methods.   *(2 marks)*

Recycling metals

Recycling metals instead of extracting them from ores has many benefits.

Extracting metals

Extracting metals from their ores:

- ✗ uses up limited resources
- ✗ uses a lot of energy
- ✗ damages the environment.

> Recycling metals reduces these disadvantages. Used metal items are collected. Rather than throwing them away, these are taken apart. The metal is melted down to make new items.

 VS

Recycling

Recycling metals means:

- ✓ metal ores will last longer
- ✓ less energy needed to recycle metals than to mine ores and extract metals
- ✓ fewer quarries and mines needed
- ✓ less noise and dust produced
- ✓ less land needed for landfill sites.

Worked example   **target D-C**

AQA SKILL
Suggest
Page 95

The flow chart shows the main stages in extracting aluminium from its ore.

Use it to suggest the benefits of recycling aluminium. *(3 marks)*

Less waste rock will be produced from mining. Aluminium oxide will not need separating and purifying from aluminium ore, which will save energy. Less carbon dioxide will be emitted because less fuel will be needed for heat and electricity and because carbon dioxide is produced from the electrolysis.

> You need to be specific in your answer. To write that recycling is 'better for the environment' does not give enough detail.

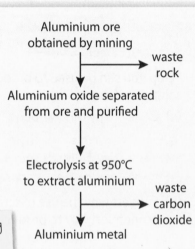

Aluminium ore obtained by mining → waste rock

↓

Aluminium oxide separated from ore and purified

↓

Electrolysis at 950°C to extract aluminium → waste carbon dioxide

↓

Aluminium metal

Recycling metals saves energy

A recent study reported these findings.

metal	% energy saving
aluminium	94
iron and steel	70
copper	86

Drawbacks of recycling

- ✗ Used metal items must be collected and transported to the recycling centre.
- ✗ Different metals must be removed from used items and sorted.
- ✗ Recycling saves different amounts of energy, depending on the metal involved.

Now try this

target D-C

1 Describe **two** ways in which recycling copper can reduce pollution.   *(2 marks)*

target B-A*

2 Suggest **two** reasons why more energy is saved when aluminium is recycled than when steel is recycled.   *(2 marks)*

3 Give **one** environmental impact and **one** ethical or social impact of recycling steel food cans.   *(2 marks)*

Steel and other alloys

Most metals in everyday use are mixtures of metals called ALLOYS.

Cast iron

Iron oxide is reduced to iron in a BLAST FURNACE. Iron straight from the blast furnace is about 96% pure. The impurities it contains make the iron BRITTLE and this limits its uses.

Blast furnace iron is used as CAST IRON.

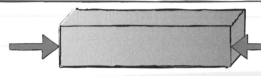

Cast iron is strong in compression. Cast iron has been used to make manhole covers, drain pipes and pillars in buildings. It is now used to make pans and garden furniture.

Steels

Most iron is converted into STEEL. There are different types of steel, but they are all alloys. Steels are mixtures of iron and carbon, often with other metals.

Low-carbon steel is easily shaped.

High-carbon steel is hard.

Stainless steel is resistant to corrosion.

Worked example   D–C

The table shows some typical properties of gold alloys.

Gold alloy	% copper	Relative strength	Relative hardness
18 carat	20.5	4.1	4.4
22 carat	5.1	2.6	2.3
24 carat	0	1	1

Notice that the gold alloys become stronger and harder the more copper they contain.

Pure gold, copper, iron and aluminium are too soft for many uses. For everyday use, they are mixed with small amounts of similar metals to make them harder. For example, copper is mixed with zinc to make brass.

(a) Use information from the table to explain why copper is mixed with gold.   (2 marks)

To make the gold stronger and harder.

(b) Suggest another reason why gold is alloyed with copper.   (1 mark)

Copper is cheaper than gold.

EXAM ALERT!

Take care to use information given in the question, rather than just repeating it.

Students have struggled with questions like this in recent exams – **be prepared!**

Now try this

1 What is an alloy?   (1 mark)

D–C

2 Explain why iron from the blast furnace has limited uses.   (2 marks)

B–A*

3 Duralumin is an alloy of aluminium, copper and magnesium. Use the data in the table to evaluate the use of duralumin and aluminium for aircraft parts.   (4 marks)

Metal	Density in g/cm³	Strength in MPa
Duralumin	2.8	450
Aluminium	2.7	150

Transition metals

The elements in the central block of the periodic table are called the TRANSITION METALS.

Transition metals

The transition metals include iron, titanium and copper.

They are good conductors of electricity.

Good for making things that need to let heat or electricity pass through them easily.

Useful as structural materials.

They are good conductors of heat.

They can be bent or hammered into shape.

The transition metals are between groups 2 and 3 in the periodic table.

Worked example B-A*

Explain why copper is used to make electrical cables and water pipes. *(2 marks)*

Copper is a good conductor of electricity, so it is useful for electricity cables. Copper does not react with water, which makes it useful for making water pipes.

Make sure the property matches the use. For example, copper is also a good conductor of heat, but this is not the relevant property for making electrical wiring and water pipes.

Copper can be bent easily, which also makes it useful for electrical cables. However, it is hard enough for pipes or tanks.

Aluminium and titanium

AQA SKILL
Identify
Page 95

Aluminium and titanium have these properties in common:
- low density (lightweight for their size)
- resistant to corrosion.

Titanium is used for artificial hip joints.

Aluminium is used for aircraft parts.

Now try this

target
D-C

target
B-A*

1 State **two** properties of copper that makes it suitable for making saucepans. *(2 marks)*

2 Aluminium and titanium resist corrosion, while steel does not. The table shows some of their other properties.

Metal	Density in g/cm³	Cost in £ per kg	Strength in MPa
aluminium	2.7	2.40	250
titanium	4.5	6.80	1000
low-carbon steel	7.8	0.85	780

(a) Name the densest metal. *(1 mark)*

(b) Explain **one** advantage and **one** disadvantage of using aluminium instead of steel for making car body panels. *(4 marks)*

(c) Compare the use of aluminium with titanium for making car body panels. *(4 marks)*

Hydrocarbons

Crude oil is a mixture of a very large number of compounds, most of which are hydrocarbons.

Alkanes

A HYDROCARBON is a compound made up of hydrogen atoms and carbon atoms *only*. Most of the hydrocarbon molecules in crude oil are ALKANES.

A **compound** consists of two or more elements chemically combined together. Take care not to write that hydrocarbons are mixtures of hydrogen and carbon.

Alkanes have the general formula C_nH_{2n+2}. For example, the chemical formula for butane (which contains four carbon atoms) is C_4H_{10}.

EXAM ALERT!

Take care when writing the formulae for alkanes: C^4H^{10} or C4H10 would be wrong.

Students have struggled with this topic in recent exams – **be prepared!**

Formulae of alkanes

An alkane molecule can be represented by its chemical formula or by its DISPLAYED STRUCTURE.

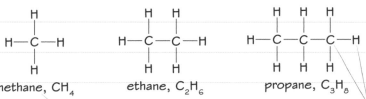

methane, CH_4 ethane, C_2H_6 propane, C_3H_8

The atoms in hydrocarbon molecules are joined together by **covalent bonds**.

The names of alkanes end in 'ane'.

Each hydrogen atom has one bond and each carbon atom has four bonds.

Worked example   B-A*

Butane is an alkane with four carbon atoms. Draw its displayed structure.   *(2 marks)*

You need to know the names of the four alkanes on this page.

When a structure is drawn out like this one, each line represents a covalent bond. Carbon atoms each form four covalent bonds, and hydrogen atoms each form only one bond. Remember that covalent bonds form when atoms share electrons.

Alkanes are saturated hydrocarbons – their carbon atoms are joined together by single covalent bonds only.

Now try this

 target D-C

1 Crude oil contains saturated hydrocarbons. State what is meant by:
   (a) saturated   *(1 mark)*   (b) hydrocarbon.   *(1 mark)*
2 State the general formula for alkanes. Use n for the number of carbon atoms.   *(1 mark)*
3 Compound X has the formula C_5H_{12}. State why X is an alkane.   *(1 mark)*

 target B-A*

4 Hexane is an alkane with six carbon atoms. Draw its displayed structure.   *(2 marks)*

Crude oil and alkanes

Mixtures

The substances in a mixture are not chemically combined together. These substances can be:

- two or more elements
- two or more compounds
- elements and compounds.

Mixing does not change the chemical properties of each substance in a mixture.

Distillation

DISTILLATION is one of several physical methods that can be used to separate the substances in a mixture. It is used to separate a mixture of liquids that have dissolved into each other. The mixture is heated until one of the liquids evaporates. Its vapours are then cooled and condensed to form a separated liquid.

Fractional distillation

FRACTIONAL DISTILLATION is used to separate mixtures containing several different substances. It is used to separate crude oil into FRACTIONS in a continuous process.

Each fraction contains molecules with a similar number of carbon atoms and boiling point.

Fractional distillation of crude oil happens in a fractionating column:

- The oil is heated to evaporate it.
- Vapour from the oil rises up the column.
- Each fraction condenses at a different temperature because it has a diferent range of boiling points.

Worked example   D-C

The diagram shows the formula and boiling point of some alkanes in three fractions.

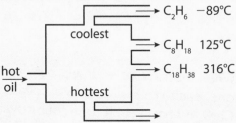

C_2H_6 −89°C

coolest

C_8H_{18} 125°C

$C_{18}H_{38}$ 316°C

hot oil →

hottest

(a) How does the number of carbon atoms in an alkane affect its boiling point?   *(1 mark)*

The higher the number of carbon atoms in a molecule of alkane, the higher the boiling point.

(b) Explain which of these alkanes is the most flammable.   *(2 marks)*

C_2H_6, because it has the smallest molecules.

There are trends in the properties of the different fractions from crude oil.

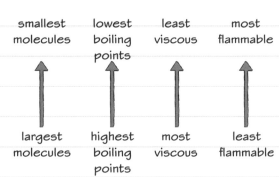

smallest molecules	lowest boiling points	least viscous	most flammable
↑	↑	↑	↑
largest molecules	highest boiling points	most viscous	least flammable

These properties influence how the fractions are used as fuels.

Now try this

 D-C

1 State the property of crude oil fractions that allows them to be separated during fractional distillation.   *(1 mark)*

2 Describe how crude oil is separated by fractional distillation.   *(3 marks)*

 B-A*

3 Explain why hydrocarbons with small molecules are better as fuels than hydrocarbons with large molecules.   *(2 marks)*

Combustion

Complete combustion

The combustion of hydrocarbon fuels releases energy to the surroundings. The hydrogen and carbon in the fuel reacts with oxygen and is OXIDISED. If there is plenty of oxygen, COMPLETE COMBUSTION occurs. The hydrogen atoms in the fuel are oxidised to produce water vapour:

$$2H_2 + O_2 \rightarrow 2H_2O$$

The carbon is oxidised to carbon dioxide:

$$C + O_2 \rightarrow CO_2$$

Partial combustion

If there is not enough oxygen, INCOMPLETE COMBUSTION (partial combustion) occurs. The hydrogen in the fuel is still oxidised to water vapour, but the carbon is not fully oxidised. These products are also formed:

- carbon monoxide
- particulates (solid particles).

The solid particles contain soot, which is carbon, and unburned fuel.

Worked example

 D-C

 Describe   Page 95

The diagram shows some of the substances released when fossil fuels are burned.

water vapour
carbon dioxide
sulfur dioxide
particulates

Describe an environmental problem caused by particulates.   *(2 marks)*

Particulates reflect sunlight back into space, causing global dimming.

Carbon dioxide is a cause of **global warming**, which leads to rising sea levels and climate change.

NO$_x$

NO$_x$ are OXIDES OF NITROGEN such as NO and NO$_2$. They are produced from nitrogen and oxygen in the air. They form at high temperatures, like those found in furnaces and car engines. NO$_x$ are also a cause of acid rain.

Sulfur dioxide

Most fuels naturally contain some sulfur. When the fuel burns, the sulfur oxidises to sulfur dioxide gas:

$$S + O_2 \rightarrow SO_2$$

Acid rain damages rocks, buildings, trees and aquatic life.

Sulfur dioxide pollution can be reduced by:

- Removing sulfur from the fuel.
- Removing sulfur dioxide after burning.

Now try this

 D-C

1 The table shows the products of combustion of two fuels, A and B.

Fuel	Carbon dioxide	Carbon monoxide	Water vapour	Sulfur dioxide
A	✓	✗	✓	✓
B	✗	✓	✗	✓

(a) Explain which fuel was a hydrocarbon.   *(2 marks)*

(b) Explain which fuel underwent incomplete combustion.   *(2 marks)*

 B-A*

2 Complete this equation for the complete combustion of propane.   *(2 marks)*

$$C_3H_8 + \ldots O_2 \rightarrow \ldots CO_2 + \ldots H_2O$$

Biofuels

BIOFUELS are produced from plant material rather than from fossil fuels.

Biofuels

Burning biofuels releases less carbon dioxide overall than burning fossil fuels.

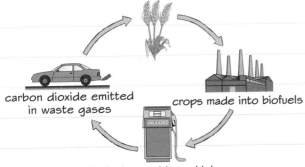

carbon dioxide absorbed in photosynthesis

carbon dioxide emitted in waste gases

crops made into biofuels

biofuels used by vehicles

Biodiesel and bioethanol

BIODIESEL and BIOETHANOL are biofuels.

- ✓ Biodiesel can be used in diesel engines.
- ✓ Bioethanol can be mixed with petrol and used in petrol engines.
- ✗ Farmland that could be used for food production is used for biofuel production instead.

Both fuels are produced from renewable resources. However, non-renewable resources may be used indirectly:

- ✗ to make fertilisers for the plant crops
- ✗ to provide energy during their manufacture and transport.

Worked example   target D-C

AQA SKILL
Evaluate
Page 95

The chart shows car exhaust emissions using diesel from crude oil, and using biodiesel.

Which fuel is better for the environment? Use your knowledge and the information above to help you answer the question.   (3 marks)

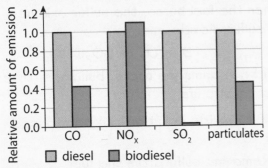

I think that generally biodiesel is better for the environment. Apart from the emissions of NO_x, emissions from biodiesel are lower than those from the ordinary diesel. The SO_2 emissions are reduced a lot more than the others. The reduction in particulates will reduce global dimming. Biodiesel releases less carbon dioxide overall than ordinary diesel does.

Hydrogen as a fuel

Water vapour is the only product made when hydrogen burns:

$$2H_2 + O_2 \rightarrow 2H_2O$$

- ✓ No carbon dioxide is made when hydrogen burns.
- ✓ Hydrogen can be made by passing electricity through water.
- ✗ Most hydrogen is produced from fossil fuels at the moment.
- ✗ Electricity needed to make hydrogen from water may be generated by burning fossil fuels.
- ✗ Hydrogen is difficult to store and there are only a few places that sell it.

You may also be asked about the social and economic impacts of using fuels.

Now try this   target D-C

1. Compare the products made when ethanol and hydrogen burn.   (2 marks)

2. Biofuel crops are valuable to farmers but farmland is needed to grow them. Suggest **one** possible problem for the food supply if the use of biofuels increases.   (1 mark)

target B-A*

3. Complete this equation for the complete combustion of ethanol.
$C_2H_5OH + \ldots O_2 \rightarrow \ldots CO_2 + \ldots H_2O$ (2 marks)

When balancing oxygen atoms remember that ethanol contains oxygen.

Chemistry six mark question 2

There will be one 6 mark question on your exam paper which will be marked for *quality of written communication as well as scientific knowledge*. This means that you need to apply your scientific knowledge, present your answer in a logical and organised way, and make sure that your spelling, grammar and punctuation are as good as you can make them.

Worked example

Metals such as steel, copper and aluminium are made into many useful things, such as parts for cars. The raw materials for these metals are extracted from metal ores.

When they are no longer needed, items containing metals may be thrown away as waste, or they may be recycled.

Describe why it is important to recycle these metals rather than disposing of them in landfill sites. *(6 marks)*

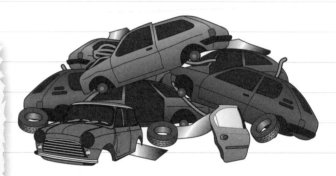

If we just throw away metal we will have to get more for the things we use. Metals are extracted from ores, which involves digging up and processing large amounts of rock. This produces a lot of waste and damages habitats. A lot of energy is needed to extract metals from their ores. For example, electricity is used to extract aluminium and to purify copper. This is expensive.

Recycling metals instead of mining for new ones means the ores will last longer. Fewer mines will be needed, which will reduce waste and loss of habitat. Fewer landfill sites will be needed, as there will be less waste. Less energy is needed to melt down a metal for recycling than is needed to extract it from its ore.

EXAM ALERT!

Make sure you plan your answer. This question asks you to about both recycling and landfill. Make sure that your answer contains information on both.

Students have struggled with questions like this in recent exams – **be prepared!**

This answer describes two environmental impacts of mining, followed by an economic impact of producing metals from their ores. The release of harmful gases might also have been mentioned. For example, carbon dioxide is released from the blast furnace when making iron. Generating electricity from fossil fuels also releases carbon dioxide.

Now try this

Aluminium and copper are both useful metals but they are extracted from their ores in different ways. Describe how aluminium and copper are extracted from their ores. *(6 marks)*

Think about how aluminium is extracted from aluminium oxide, and how copper is extracted from copper oxide. What steps are involved in each process?

Cracking and alkenes

Cracking is a reaction in which hydrocarbons are broken down to form smaller molecules, including alkanes and ALKENES. Some of these products are useful as fuels.

Cracking

CRACKING is a THERMAL DECOMPOSITION reaction. In cracking, oil fractions are heated so they vaporise. Their vapours are either:

- passed over a hot catalyst, or
- mixed with steam and heated to very high temperatures.

Alkenes

Alkenes are hydrocarbons. They are UNSATURATED because their molecules contain one or more carbon to carbon double bonds.

Alkenes have the general formula C_nH_{2n}. For example, the chemical formula for butene (which contains four carbon atoms) is C_4H_8.

Formulae of alkenes

An alkene molecule can be represented by its chemical formula or by its displayed structure.

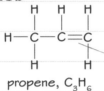

propene, C_3H_6

The = represents a double bond. The C=C bond allows alkenes to react with BROMINE WATER. Alkenes turn it from orange to colourless (alkanes cannot do this). Bromine water can be used as a test for unsaturated molecules.

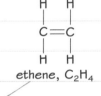

ethene, C_2H_4

Alkenes – double e, double bond

Worked example   B-A*

(a) Complete the balanced equation for the cracking of the hydrocarbon C_6H_{14} to produce ethene and an alkane.   *(2 marks)*

$$C_6H_{14} \rightarrow C_2H_4 + C_4H_{10}$$

(b) Suggest **two** reasons why there is a greater demand for the products than for the original hydrocarbon.   *(2 marks)*

Smaller hydrocarbons make better fuels than larger ones. Alkenes are used to make polymers.

Remember:
- alkanes – the number of H atoms is double the number of C atoms plus two
- alkenes – the number of H atoms is double the number of C atoms.

This reaction is also possible:
$$C_6H_{14} \rightarrow 2C_2H_4 + C_2H_6$$

Compared with larger hydrocarbons, smaller hydrocarbons are less viscous (more runny), more flammable and have lower boiling points (they tend to be gas or liquid). They are more useful as fuels.

Now try this

B-A*

1 The diagram shows an experiment to crack paraffin soaked onto mineral wool.

(a) Describe a test for the presence of alkenes.   *(2 marks)*

(b) C_2H_4 molecules are produced. State whether they would collect at position X or position Y.   *(1 mark)*

(c) Draw the displayed structure of propene, C_3H_6.   *(1 mark)*

(d) Describe **two** reasons for cracking.   *(2 marks)*

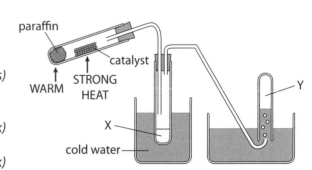

Making polymers

Alkenes can be used to make polymers (plastics).

Polymerisation

Alkene molecules can act as MONOMERS. They can join together in a POLYMERISATION reaction to make very large molecules called POLYMERS.

For example, ethene forms poly(ethene).

monomers

polymer

$$n \; C=C \longrightarrow \left(\begin{array}{cc} H & H \\ | & | \\ C & C \\ | & | \\ H & H \end{array} \right)_n$$

The displayed structures show that many (n) ethene monomers can react together to form n units of poly(ethene).

(a) Draw the displayed structure of the polymer that forms from propene. *(1 mark)*

$$n \; C=C \longrightarrow \left(\begin{array}{cc} H & H \\ | & | \\ C & C \\ | & | \\ H & CH_3 \end{array} \right)_n$$

(b) Name the polymer formed. *(1 mark)*

Poly(propene)

The name of a polymer is given by its monomer – it is poly(name of monomer).

EXAM ALERT!

You should be able to show the formation of a polymer from a given alkene monomer. Students have often struggled to do this in exams. They often forget to change the double bond to a single bond or they leave out the n.

Students have struggled with questions like this in recent exams – **be prepared!**

To convert from a monomer to a polymer:
- draw the monomer but with a single bond
- draw a long bond either side
- draw brackets through the long bonds
- write n after the bracket.

Uses of polymers

New polymers and uses of polymers are being developed. Polymers developed for one use may be used in new ways.

Polymers are being developed as medical dressings and shape memory polymers as well as:

hydrogels (for babies nappies)

false teeth and fillings

---

target **D-C**

1 The inside of babies' disposable nappies contains a powdered hydrogel underneath a lining layer. Suggest **two** properties the hydrogel should have. *(2 marks)*

2 (a) Complete this equation to show how a polymer forms from chloroethene. *(1 mark)*

$$n \; C=C \longrightarrow$$
with
$$\begin{array}{cc} H & H \\ | & | \\ C=C \\ | & | \\ H & Cl \end{array}$$

(b) Name the polymer formed. *(1 mark)*

target **B-A***

3 Explain whether poly(ethene) is a saturated or an unsaturated hydrocarbon. *(2 marks)*

Polymer problems

A BIODEGRADABLE material can be broken down by microbes. CORNSTARCH is a natural substance that microbes can break down. It is used for making biodegradable substances such as carrier bags.

But many polymers are not biodegradable. Microbes cannot break them down and they do not rot.

This is a useful property because items made from polymers last a long time and may be recycled.

Polymers that are not biodegradable are difficult to dispose of – sometimes they cause litter.

Landfill sites

Most waste goes into LANDFILL:

- ✓ Waste is disposed quickly.
- ✓ Waste is out of sight once it is covered over.
- ✗ Space for landfill sites is running out.
- ✗ Most polymers are not biodegradable and will last for many years.
- ✗ Landfill sites are unsightly and attract pests.

Worked example

 target B-A*

 AQA SKILL Consider Page 95

Give **two** advantages and **two** disadvantages of the disposal of polymers by recycling rather than by landfill. *(4 marks)*

Dumping polymers in landfill sites is a waste of a non-renewable resource, as polymers are made from crude oil. Recycling means that less waste goes into landfill. On the other hand, recycling is expensive because the different polymers must be collected and sorted. Some polymers cannot be recycled.

Worked example

 target D-C

Suggest why it has become worthwhile for companies in China to pay for waste polythene to be shipped to them from America and Europe. *(2 marks)*

Polythene is made from crude oil. The cost of crude oil may have risen to the point where it is cheaper to recycle polythene instead.

Fuels vs. polymers

Most crude oil is used for fuel, but it is also used to make chemicals. These include medicines and paints, as well as polymers. Oil is a limited resource. It will run out one day if we keep using it.

This answer looks at the economic aspects of recycling polymers, but there are environmental and social aspects too. Recycling creates new jobs, for example in the transport industry.

Now try this

 target D-C

1 Suggest **one** reason why it may be better to recycle polymers, rather than to dispose of them in landfill sites. *(2 marks)*

 target B-A*

2 The diagram shows the displayed structure for polylactic acid, a biodegradable polymer.

(a) Explain whether or not polylactic acid is a hydrocarbon. *(2 marks)*

(b) State what the term biodegradable means. *(1 mark)*

(c) Give **one** advantage of making plastic bags from cornstarch. *(1 mark)*

Ethanol

Ethanol is the alcohol in alcoholic drinks. It is also a fuel, and it has industrial and chemical uses. There are two ways to produce ethanol:

Fermentation

FERMENTATION is a natural process. It uses yeast (a type of microorganism) to convert sugar from plants into ethanol:

sugar → ethanol + carbon dioxide

$$C_6H_{12}O_6 \rightarrow 2C_2H_5OH + 2CO_2$$

Fermentation is used to make alcoholic drinks and most of the ethanol used as a fuel.

Hydration of ethene

Most ethanol for industrial use is made by hydration of ethene. Ethene and steam react together in the presence of a catalyst to make ethanol:

$$C_2H_4 + H_2O \rightarrow C_2H_5OH$$

Worked example   D–C

AQA SKILL
Explain
Page 95

The information below is about the two ways to produce ethanol. Explain which one would be better to use in a country with lots of land but not a lot of crude oil. *(2 marks)*

Fermentation
- Raw material (sugar) is a renewable resource.
- Reaction is slow but does not need very high temperatures.
- Ethanol must be purified before use.
- Land to grow the plants needed could be used instead for food crops.

Hydration of ethene
- Raw material (crude oil) is non-renewable.
- Reaction is fast but needs high temperatures.
- Pure ethanol is made.
- Ethene is made by fractional distillation of oil, and then cracking. Both processes need energy.

Fermentation would be better in this case. A country like this would have lots of land to grow plants for sugar, but crude oil would be too expensive to use to make ethene for ethanol.

Now try this

target
D–C

1 Suggest why most ethanol for use as fuel is produced by fermentation, rather than by hydration of ethene. *(2 marks)*

2 Describe what is seen during fermentation. *(1 mark)*

target
B–A*

3 Draw the displayed structures for the molecules in the reaction between ethene and steam. *(2 marks)*

4 Ethanol can be produced from sugar, or from crude oil. Describe **one** advantage and **one** disadvantage of each method. *(4 marks)*

Vegetable oils

There are two ways that oils can be extracted from seeds, nuts and fruits:

Pressing and filtering

The plant material must be crushed to release oil and water from the plant cells. The oil floats on top of the water and crushed plants.

vegetable oil
water
skin and seeds

The crushed plant material may need PRESSING to remove the oil. The crushed plant material can be removed by FILTRATION. The oil and water are separated to make a useful product.

Steam distillation

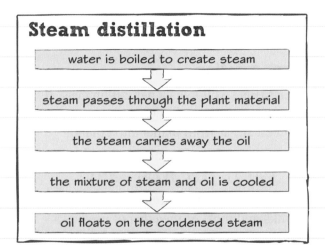

water is boiled to create steam

↓

steam passes through the plant material

↓

the steam carries away the oil

↓

the mixture of steam and oil is cooled

↓

oil floats on the condensed steam

Worked example   target **D-C**

AQA SKILL
Explain
Page 95

The table shows the amount of energy in 100 g of three foods.

Food	Energy in kJ per 100 g
boiled potato	360
fried potato wedges	1180
potato crisps	2240

(a) Explain why the wedges and crisps contain more energy than the boiled potato.   *(2 marks)*

They are cooked in oil, which increases the energy the food releases when eaten.

(b) Suggest a possible health effect of eating crisps rather than boiled potato.   *(1 mark)*

You could become overweight.

Cooking with oils

Food can be cooked by boiling in water. Vegetable oils have higher boiling points than water. This means that food cooked in oil:

✓ cooks at higher temperatures

✓ cooks faster than in water

✓ has different flavours

✗ releases more energy when it is eaten.

In an exam you may be given data to evaluate. This usually means that you have to know some advantages and disadvantages. Vegetable oils are important foods because they provide us with nutrients and a lot of energy. However, too much can make us overweight, and this can cause health problems.

Now try this

target **D-C**

1 Describe **two** reasons why sunflower oil, not water, is used to cook potato for crisps.
   (2 marks)

target **B-A***

2 Lavender oil may be separated from crushed lavender by steam distillation. The diagram shows how this may be done.
   Describe how lavender oil is produced this way.
   (4 marks)

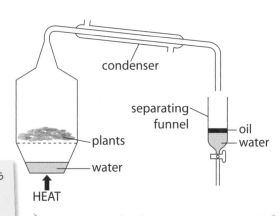

condenser
separating funnel
oil
water
plants
water
HEAT

Describe the process, starting with what happens when the water is heated, and continuing to where oil floats on water in the funnel.

Emulsions

Oils do not dissolve in water. If an oil and water are shaken together, they form a mixture called an EMULSION.

Compared with oil or water alone, emulsions:

- ✓ are thicker (more viscous)
- ✓ have better coating ability (they stick to food or other objects better)
- ✓ have a better texture and appearance.

This makes them useful for paints, salad dressings, ice cream and cosmetics.

Emulsifiers

An emulsion will eventually separate out again until all the oil is floating on the water. EMULSIFIERS are substances that make emulsions more STABLE (they do not separate out after mixing).

For example, the natural emulsifiers in egg yolk prevent the vinegar and oil in mayonnaise separating out after mixing.

Worked example　target B-A*

AQA SKILL
Consider
Page 95

Give **one** advantage and **one** disadvantage of using egg lecithin in food.　*(2 marks)*

An emulsifier like egg lecithin stops oil and water separating. This increases the shelf-life of the food. However, there is a possible risk to people who are allergic to eggs.

Emulsifiers are important ingredients for making emulsions stable. They are included in the list of ingredients to let people make informed decisions about whether to eat the food.

When given information about emulsifiers in foods, you should be able to evaluate their advantages, disadvantages and risks.

Emulsifier molecules

Emulsifier molecules have two parts.

hydrophilic 'head'　hydrophobic 'tail'

dissolves in water　dissolves in oil

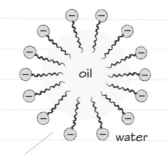
oil
water

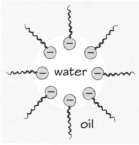
water
oil

'Hydrophilic' means water loving and 'hydrophobic' means water hating.

Emulsifier molecules surround droplets of oil or water, stopping them joining up again. This works whether the emulsion is made of oil droplets in water, or water droplets in oil.

Now try this

target D-C

1　Explain why emulsions are useful in household paint.　*(2 marks)*

2　A simple salad dressing can be made by shaking olive oil with vinegar.
　(a) Explain why a layer of olive oil forms on top of the vinegar after a while.　*(2 marks)*
　(b) Suggest why egg yolk is an ingredient in many salad dressing recipes.　*(2 marks)*

target B-A*

3　(a) Describe the structure of an emulsifier molecule.　*(2 marks)*
　(b) Explain how an emulsifier is able to stabilise an emulsion of water droplets in oil.　*(3 marks)*

Hardening plant oils

Unsaturated vegetable oils

Vegetable oil molecules have long chains of carbon atoms joined together:

- Saturated vegetable oils only have carbon to carbon single bonds, C–C.
- UNSATURATED vegetable oils have some carbon to carbon double bonds, C=C, as well.
- POLYUNSATURATED oils have lots of C=C bonds. Unsaturated oils are better for health than saturated oils and fats.

Testing for unsaturation

Bromine water is orange. It turns colourless when it is mixed with an unsaturated substance. The C=C bonds change to C–C bonds in the reaction. This means:

- the more C=C bonds there are, the more bromine water can be added before it stops becoming colourless.

Worked example   target **D-C**

Bromine water was added to an equal mass of each oil until it stayed orange.

Vegetable oil	Volume of bromine water in cm³
Olive oil	15.6
Corn oil	24.8
Cocoa butter	7.4

(a) What would you see when the first drops of bromine water are added to an oil?   *(2 marks)*

The bromine water turns from orange to colourless.

(b) What do the results show you?   *(2 marks)*

Olive oil is more unsaturated than cocoa butter but less unsaturated than corn oil.

EXAM ALERT!

Do not write 'clear' when you mean 'colourless'! A substance can be coloured but still clear.

Students have struggled with questions like this in recent exams – **be prepared!**

Hardening vegetable oils

Unsaturated oils can be HARDENED.

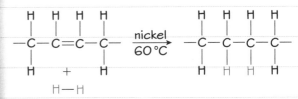

A nickel catalyst speeds up the reaction between hydrogen and the C=C bonds.

Hardened vegetable oils are described as HYDROGENATED. They are less unsaturated than the oils used to make them.

Hydrogenated oils have higher melting points than the original oils. They are:

- solid at room temperature
- used to make margarine
- useful in cakes and pastries.

Now try this

target **D-C**

1 Describe what happens when bromine water is shaken with an unsaturated oil.   *(2 marks)*

target **B-A***

2 Margarine may contain hardened vegetable oils.
  (a) Describe how vegetable oils are hardened.   *(3 marks)*
  (b) Explain why vegetable oils are hardened.   *(3 marks)*

The Earth's structure

The Earth has a layered structure.

Three main layers

The Earth is surrounded by an ATMOSPHERE and has three main layers.

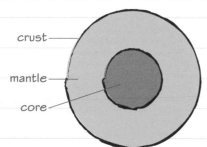

The CRUST is the thinnest layer. The MANTLE is a thick layer between the crust and the core. The radius of the CORE is just over half the Earth's radius.

Tectonic plates

The Earth's crust and the upper part of the mantle are cracked into several very large pieces. These are called TECTONIC PLATES.

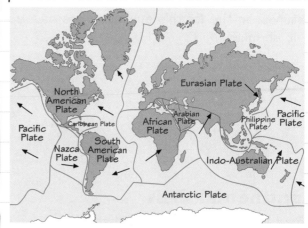

Moving plates

The tectonic plates move at a few centimetres per year. They move because of CONVECTION CURRENTS in the mantle. These currents are driven by HEAT from natural RADIOACTIVE processes in the Earth.

The mantle is mostly solid but it can move slowly. Tectonic plates can move towards or away from each other. They can also move past each other. Most **earthquakes** and **volcanoes** happen at the boundaries between plates.

Now try this

 target D-C
1 Describe the Earth's layered structure. *(3 marks)*
2 Describe why tectonic plates move. *(2 marks)*

 target B-A*
3 Read this newspaper report from 2012.
   (a) Describe how earthquakes are caused. *(2 marks)*
   (b) Describe the problems of trying to predict earthquakes accurately. *(2 marks)*

Italian scientists jailed for six years
An Italian court has found six scientists guilty of negligence. It said they failed to give adequate warning of an earthquake in 2009 that measured 6.3 on the Richter Scale. The city of L'Aquila was badly damaged and hundreds of people were killed or injured.

Continental drift

Alfred Wegener was a German scientist who proposed a theory of crustal movement, called CONTINENTAL DRIFT. It was not generally accepted for many years.

Shrinking Earth

Before Wegener, scientists thought that features on the Earth's surface were caused by it shrinking.

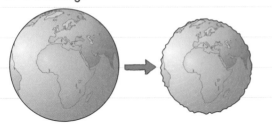

The idea was that as the Earth cooled after it was formed, its crust shrank and wrinkled to form mountains. The idea is now disproved.

Prediction: If this theory is correct then mountains should cover the Earth.

Observation: mountains are not everywhere.

Continental drift

Alfred Wegener suggested that all the Earth's land was once joined together, forming a 'supercontinent'. This broke up millions of years ago and the landmasses moved apart.

Wegener's ideas were based on observations involving South America and Africa.

Evidence for continental drift

Wegener's evidence for his theory included:

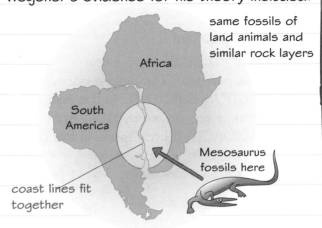

same fossils of land animals and similar rock layers

Africa

South America

Mesosaurus fossils here

coast lines fit together

Worked example   target **D-C**

AQA SKILL
Identify
Page 95

When Wegener proposed his theory of continental drift in 1912, other scientists thought that he was wrong. Give **one** reason why they thought this.   *(1 mark)*

Wegener had no evidence to show how continents could move.

Wegener was not a geologist. He had evidence to support his idea but couldn't explain how continents moved. Evidence that the continents are parts of moving tectonic plates was not discovered until much later.

Now try this

target **D-C**

1 Describe what Alfred Wegener meant by 'continental drift'.   *(2 marks)*

2 Suggest **two** reasons why other scientists initially thought that Wegener's idea was wrong.   *(2 marks)*

target **B-A***

3 Calculate how quickly South America and Africa are moving apart. Assume that they began to separate 140 million years ago and that the average width of the South Atlantic Ocean is 2800 km.   *(2 marks)*

The Earth's atmosphere

The Earth's ATMOSPHERE has stayed much the same for the last 200 million years.

The atmosphere

The two main gases in the atmosphere are:
- nitrogen (about 4/5ths)
- oxygen (about 1/5th).

There are smaller amounts of other gases.

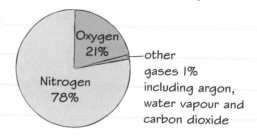

Oxygen 21%
Nitrogen 78%
other gases 1% including argon, water vapour and carbon dioxide

---

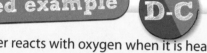

Worked example   target **D-C**

Copper reacts with oxygen when it is heated. In an experiment, a sample of air was repeatedly passed over hot copper turnings.

The volume of air at the start was 50 cm³. At the end, after the apparatus had cooled, the volume was 40 cm³. Calculate the percentage of oxygen in the air.   *(2 marks)*

volume of oxygen = 50 − 40 = 10 cm³
percentage of oxygen = 10/50 × 100
                     = 20%

Raw materials

Earth's atmosphere, crust and oceans are the only sources of the minerals and other resources needed by humans. For example, air is the raw material for producing oxygen for industrial and medical use.

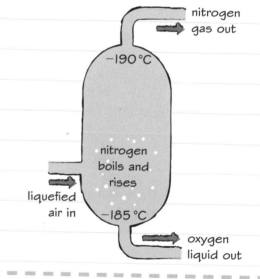

nitrogen gas out
−190 °C
nitrogen boils and rises
liquefied air in   −185 °C
oxygen liquid out

Air as a raw material

Gases in air can be used as raw materials for industrial processes. The gases have different boiling points, so they can be separated by FRACTIONAL DISTILLATION:
- air is cooled to −200 °C and liquefied
- liquid air goes into a fractionating column
- nitrogen gas leaves at the top
- liquid oxygen leaves at the bottom.

Noble gases like argon are also separated.

Now try this

target **D-C**

1 In an experiment to find the percentage of oxygen in air, 100 cm³ of air was repeatedly passed over hot copper turnings.
   (a) Balance this symbol equation: ... Cu + O₂ → ... CuO   *(1 mark)*
   (b) The volume at the end was 79 cm³. Calculate the percentage of oxygen.   *(2 marks)*

target **B-A***

2 Air can be fractionally distilled. Use data from the diagram and the table to answer the questions.

Gas	carbon dioxide	oxygen	nitrogen
boiling point in °C	−78	−183	−196
freezing point in °C	−78	−219	−210

   (a) Suggest why carbon dioxide is removed before the air is cooled to −200 °C.   *(2 marks)*
   (b) Explain why nitrogen leaves the column as a gas but oxygen leaves as a liquid.   *(4 marks)*

The early atmosphere and life

There are different theories about how the Earth's atmosphere and life formed and developed.

| Earth's early atmosphere | = | Like Venus and Mars today: mainly carbon dioxide with little or no oxygen | + | Water vapour, small amounts of ammonia and methane |

According to one theory, the Earth's early atmosphere was similar to the atmospheres of Venus and Mars today. However, the Earth also has oceans, which formed when water vapour condensed.

Worked example

 target B-A*

 AQA SKILL Explain Page 95

Stanley Miller and Harold Urey carried out experiments in 1952. They wanted to see if amino acids could have formed on the early Earth. They sealed a mixture of water vapour, ammonia, methane and hydrogen in a sterile flask. Amino acids were found after a week. Explain how each part of the experiment was intended to simulate the conditions thought to exist on the early Earth. *(3 marks)*

The water represented the oceans and the gases were chosen to represent the atmosphere thought to have existed then. The sparks represented lightning.

This is just one of the many theories to explain how life formed on Earth billions of years ago.

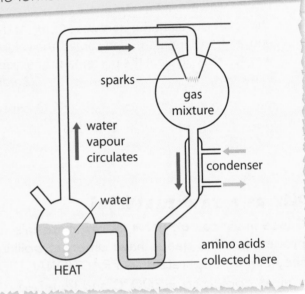

Water is H$_2$O, ammonia is NH$_3$ and methane is CH$_4$. These compounds contain the same elements as amino acids. Amino acids are essential for life, as they form proteins.

Many theories

The Miller–Urey experiment shows how a PRIMORDIAL SOUP could have formed on Earth. Primordial soup is a mixture of different molecules that could then have started to build into larger molecules and eventually to form cells. The experiment does not prove that things did happen that way. We do not know how life was first formed as it happened a very long time ago.

Now try this

target D-C

1 Describe **two** differences between the atmospheres of Earth, and Venus and Mars today. *(2 marks)*

target B-A*

2 (a) Outline the 'primordial soup' theory about how life was formed. *(3 marks)*

(b) Suggest **one** reason why we do not know how life was first formed. *(1 mark)*

Gas	% of atmosphere today	
	Venus	Mars
nitrogen	3.5	2.7
oxygen	trace	0.1
carbon dioxide	96.5	95.3

Evolution of the atmosphere

Carbon dioxide levels

As the Earth's atmosphere has developed

- carbon dioxide levels have gone down
- oxygen levels have gone up.

Photosynthesis by plants and algae is one reason for these changes.

In photosynthesis ...
- carbon dioxide in
- oxygen out

Fossil fuels

FOSSIL FUELS formed over millions of years from the remains of dead plants and animals. The carbon they contain originally came from the atmosphere when the organisms were alive. Carbon dioxide was 'locked up' in fossil fuels as:

- carbon in coal
- hydrocarbons in oil and gas.

Coal is a type of SEDIMENTARY ROCK.

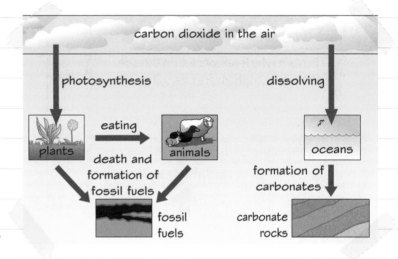

carbon dioxide in the air

photosynthesis — dissolving

plants — eating — animals — oceans

death and formation of fossil fuels

formation of carbonates

fossil fuels

carbonate rocks

Worked example   target D-C

The oceans have an important part to play in absorbing carbon dioxide from the atmosphere. Describe the processes involved.   *(3 marks)*

Carbon dioxide dissolves in the oceans. It forms carbonates, which sea creatures use to make their shells and skeletons. These form limestone rock when the animals die and their shells sink to the seabed. In this way, carbon dioxide becomes locked in sedimentary rocks as carbonates as well as fossil fuels.

Now try this

target D-C
target B-A*

1   Explain how the Earth's atmosphere has changed because of the evolution of plants and algae.   *(3 marks)*

2   Use the data in the table to help you answer the questions.

	Temperature in °C
Boiling point of water	100
Average surface temperature on Venus	460
Average surface temperature on Earth	14

(a) Suggest why Venus has no oceans today.   *(1 mark)*

(b) The Earth's oceans formed around 250 million years after the Earth itself formed. Suggest **one** reason for this.   *(2 marks)*

(c) Suggest **two** reasons why the percentage of carbon dioxide in the atmosphere decreased gradually over 3 billion years.   *(2 marks)*

Carbon dioxide today

Human activities are releasing gases into the atmosphere, causing changes to its composition.

Carbon dioxide in the oceans

Carbon dioxide is SOLUBLE in water. This is why carbon dioxide from the atmosphere DISSOLVES in the oceans. The oceans contain a lot of water, so they act as a huge RESERVOIR for carbon dioxide.

Carbon dioxide dissolves in water to form an ACIDIC solution. When the oceans absorb carbon dioxide, the pH of the water is reduced (it becomes more acidic). This affects the environment in the oceans.

Worked example **D-C**

AQA SKILL **Explain** *Page 95*

Explain why levels of carbon dioxide in the atmosphere are changing.

Fossil fuels release carbon dioxide when they burn. The use of fossil fuels is increasing the level of carbon dioxide in the atmosphere.

Extra carbon dioxide has negative impacts:
- acidification of the oceans
- global warming.

Changing atmosphere

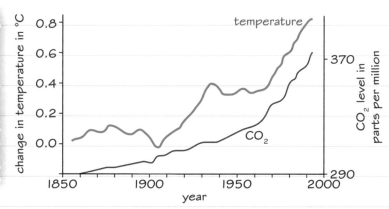

As the level of carbon dioxide in the atmosphere has risen, so has the average global temperature.

Worked example **D-C**

(a) Complete the word equation. *(1 mark)*

carbonate + acid → salt + <u>carbon dioxide</u> + <u>water</u>

(b) Calcium carbonate is the major component of seashells. Carbon dioxide increases the acidity of seawater. Suggest a problem this may cause marine organisms. *(1 mark)*

Their shells may react with the acid in the water and wear away.

Carbonates release carbon dioxide when they react with acid. Limestone is a sedimentary rock containing calcium carbonate. It is damaged by acid rain.

The oceans absorb carbon dioxide released into the atmosphere when fossil fuels are used. However, as this happens, the acidity of the water increases. This has an impact on marine organisms with shells.

Now try this

 D-C

1 (a) State why the oceans are described as being a 'reservoir' for carbon dioxide. *(1 mark)*

(b) Describe **one** reason why human activities are increasing the amount of carbon dioxide in the atmosphere. *(1 mark)*

 B-A*

2 Explain how human activities affect the amount of carbon dioxide in the atmosphere. *(2 marks)*

Chemistry six mark question 3

There will be one 6 mark question on your exam paper which will be marked for *quality of written communication* as well as *scientific knowledge*. This means that you need to apply your scientific knowledge, present your answer in a logical and organised way, and make sure that your spelling, grammar and punctuation are as good as you can make them.

Worked example

The table shows information about different hydrocarbon fractions in a sample of crude oil.

The supply is the amount of each fraction in the crude oil. The demand is the amount that can be sold. Fractional distillation of the crude oil produces too much of some fractions but too little of other fractions. Cracking can be used to reduce this problem.

Fraction	Amount in tonnes		Molecule size
	Supply	Demand	
gases	8	8	smallest
petrol	16	26	
diesel	14	20	↓
kerosene	15	9	
heavy oil	20	15	
bitumen	36	24	largest

Describe the conditions needed for cracking. Use information from the table to explain which fractions should undergo cracking. *(6 marks)*

Cracking involves heating a fraction so that its hydrocarbons vaporise. The vapours are passed over a hot catalyst or mixed with steam and heated to very high temperatures.

In cracking, large hydrocarbon molecules produce smaller, more useful hydrocarbon molecules. The table shows that the demand for petrol and diesel is greater than the supply. On the other hand, the supply of bitumen and heavy oil is greater than the demand. These fractions should be cracked to make more petrol and diesel.

The first part of the question is answered here. Make sure you answer the whole question – here it would be easy to concentrate on the table and to forget this is needed.

The answer describes what cracking does so that it becomes clear why cracking is needed. Information from the table is used to explain which fractions should be chosen. You could quote numbers but you would still need to explain what they mean.

Now try this

The Earth has a layered structure. Its surface is broken into large pieces called tectonic plates, which are constantly moving. Explain how these tectonic plates move and describe the possible results of this movement. *(6 marks)*

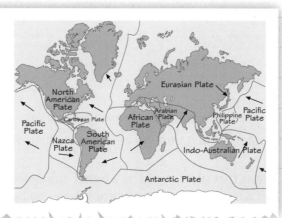

The command word EXPLAIN means you need to make something clear or state reasons for it happening.

Infrared radiation

The amount of INFRARED RADIATION emitted or absorbed by an object depends on its temperature and its surface.

All objects EMIT and ABSORB infrared radiation. The hotter an object is, the more infrared radiation it gives out in a given period of time.

> **Emit** means 'give out'; **absorb** means 'take in'.

Colours and radiation

Surface properties of object	Radiation emitter	Radiation absorber
dark, matt	good	good
light, shiny	poor	poor

Light, shiny objects are also good REFLECTORS of infrared radiation.

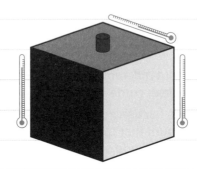

> The sides of the cube are at the same temperature but the thermometer near the dark, matt surface shows the highest temperature because the most radiation is emitted from the surface.

Worked example target D-C

AQA SKILL
Explain
Page 95

1 Explain why it might be a good idea to paint radiators in a house matt black.
(2 marks)

> The black radiators will give out more heat because black is a good emitter of infrared radiation.

EXAM ALERT!

'Explain' means to give a reason for something, so the answer usually has 'because' in it.

> Students have struggled with questions like this in recent exams – **be prepared!**

2 Police use infrared cameras to locate people in a wood who have lit a fire.

(a) Explain why the image of the fire looks brighter than the images of the people.
(2 marks)

> The fire is at a higher temperature than the people. The higher the temperature, the more energy is emitted as infrared radiation.

(b) Explain why infrared cameras are more effective at night.
(2 marks)

> Infrared radiation from the sun reflects off the ground and other objects and obscures the images of the people.

Now try this

target D-C

1 A design for a solar heater has water flowing in black tubes.

(a) State why the water passing through becomes warmer when sunlight falls on the heater.
(1 mark)

(b) Someone suggests that silver would be a better colour for the screen. Discuss whether a black or silver screen would give the best results.
(2 marks)

Kinetic theory

The kinetic theory says that objects are made up of PARTICLES that are constantly moving.

Properties of matter

The KINETIC THEORY can be used to explain the properties of SOLIDS, LIQUIDS and GASES.

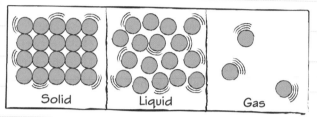

Solid   Liquid   Gas

Solids: **particles** can just **vibrate** about fixed positions. Solids have a fixed shape.

Liquids: Particles move around but stay close together. Liquids can flow.

Gases: Particles move fast and have lots of space between them. Gases fill a container and can flow.

Changes of state

When particles in a solid are given more energy they can start to move around and become a liquid.

When particles in a liquid are given more energy they move apart and become a gas.

Particles in a gas have more energy than the same particles in a liquid, which in turn have more energy than the same particles in a solid.

The amount of energy a particle has is related to the temperature of the material.

Worked example

target D-C

1   Compare the motion of particles in a solid and a gas.   (2 marks)

In a solid particles can only vibrate about a fixed position.
In gas the particles move around within their container.

target B-A*

2   Explain the following properties of solids, liquids and gases using the kinetic theory.

   (a) Solids and liquids cannot be compressed.   (2 marks)

   The particles in a solid or liquid are very close together. Solids and liquids cannot be compressed because the particles cannot get any closer to each other.

   (b) When a solid is melted, the liquid formed has a higher temperature.   (2 marks)

   The particles in the liquid are moving faster and so have more energy.

Now try this

target B-A*

1   A teacher has a box containing a number of small plastic balls. Describe how the teacher can use this equipment to demonstrate the kinetic theory for a solid, a liquid and a gas.   (6 marks)

This is not a 'six mark question' although it has 6 marks. You just need to state six valid points.

Methods of transferring energy

Energy is transferred from one place to another by CONDUCTION, CONVECTION, RADIATION, EVAPORATION and CONDENSATION.

Conduction

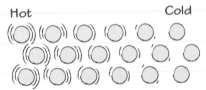

Hot   Cold

When particles in a solid are given more energy they vibrate more. They collide with neighbouring particles, which causes these particles to vibrate more. In this way energy spreads through the solid. In metals there are free electrons that help transfer the energy, making metals good conductors.

Convection

When particles in a liquid or gas are given more energy, they move faster and spread out. The density is lower so the hot material rises. It is replaced by cooler material, so a convection current is set up.

Radiation

Particles give out infrared radiation.

See page 70 for more on radiation.

Evaporation and condensation

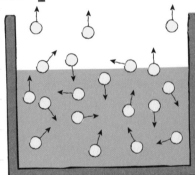

high energy: evaporating

medium energy: pulled back into water

lower energy: remain as liquid

The fastest-moving particles on the surface of a liquid escape as a gas. This is called evaporation. The average energy of the remaining particle is less, so the remaining liquid is cooler.

If energy is transferred from particles in a gas the particles will form droplets of liquid. This is called condensation.

Worked example   target B-A*

AQA SKILL
Evaluate
Page 95

1 The purpose of the radiator in a car is to transfer energy from the engine to the surroundings. Hot water flows through the radiator, which is made of copper painted black. Air flows over the surface of the radiator when the car is moving. Evaluate the design of the car radiator.   (3 marks)

The radiator transfers energy successfully because copper is a metal. Metals are good conductors so the copper will transfer energy from the hot water to the surface of the radiator. Air particles hitting the radiator gain energy and transfer energy from the radiator by convection. The black paint increases the amount of infrared radiation emitted by the radiator. The design works well.

Now try this

target D-C

1 Perfume or aftershave lotion feels cold when you put it on your skin. Explain why this happens.   (3 marks)

target B-A*

2 Polythene is a poor conductor while gold is an excellent conductor. Explain the difference in properties of the two materials.   (2 marks)

Rate of energy transfer

Energy transfer can be speeded up or slowed down by changing the materials or the design of an object.

Factors affecting energy transfer

The bigger the temperature difference between an object and its surroundings, the faster energy is transferred.

The rate at which energy is transferred to or from the object depends on:

- The material from which the object is made or the material it is packed in. Insulators transfer heat more slowly.
- The surface area and volume of the object. Greater surface area means faster rate of energy transfer.
- The properties of the surface of the object. Dull, dark surfaces emit or absorb radiation at a higher rate.

Evaporation

Evaporation takes place faster when the temperature is higher, there is a large surface area in contact with the air, or when wind blows the water vapour away.

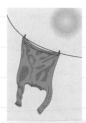

Condensation is the opposite of evaporation, so it happens faster if the temperature is low.

Polar bears

Polar bears keep warm with thick fur. The fur traps air so less energy is lost by convection. The fur is white so it radiates energy at a lower rate.

You also need to be able to apply what you know about energy transfers to buildings and how humans cope with low temperatures.

Worked example

target D-C

A student noticed that tea would keep warm in a vacuum flask for several hours.

Evaluate the methods used in the flask to reduce the rate of energy transfer.   (3 marks)

The flask is made of an insulating material, which slows the energy transfer from the hot liquid to the outside by conduction. There is a vacuum between the two layers so energy cannot be transferred by convection. The silver coating reduces the amount of energy transferred by radiation. All of this means that the flask will keep drinks at a high temperature.

- plastic cap
- vacuum
- glass with silvered surfaces
- inner silvered surface
- outer silvered surface
- outer case
- foam plastic support

Now try this

target D-C

1 Explain why a drink cools more quickly when you blow on the surface.   (3 marks)

target B-A*

2 There is a correlation between the size of the ears of varieties of fox and the average temperature of their habitat. Suggest why this is not a chance correlation.   (2 marks)

Keeping warm

Radiators are one way of warming a house up, but once it is warm then it is important that the house is insulated or the energy will be transferred to the surroundings and the house will cool down again. U-values are useful in choosing how to insulate your home.

Worked example

D-C

AQA SKILL Evaluate Page 95

The diagram shows a metal radiator that transfers 600 J of energy to the surroundings every second. Evaluate features of the design that affect the transfer of energy to the surroundings. *(3 marks)*

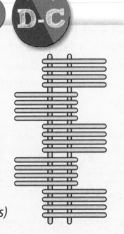

The radiator is made of a metal that is a good conductor. It has a large surface area that increases convection. However, it is coloured silver, which reduces the radiation of energy. The design is generally good but could be improved.

U-values

Energy is transferred through a material when the temperature on each side is different.

U-values compare how much energy is transferred through materials in a given time. The values are for a given thickness and area of material.

The lower the U-value, the better the material is as an insulator. Metals have high U-values. Wool and foams have low U-values.

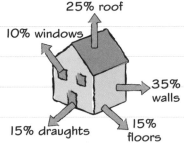

25% roof
10% windows
35% walls
15% draughts
15% floors

This picture shows how energy is usually lost from a house. Using materials with lower U-values can cut the amount of energy transferred.

Worked example

B-A*

AQA SKILL Evaluate Page 95

See page 76 for more details about payback time.

Use the information in the table to answer the questions.

Material	U-value (10 cm thick)	Cost to cover loft in £	Reduction in energy bills per year in £	Payback time in years
sheepwool	0.33	230	100	2.3
rockwool	0.20	380	190	2

The table gives data on materials used to insulate lofts. Evaluate the cost effectiveness of each of the materials. *(3 marks)*

Rockwool is the better insulator because it has the lowest U-value, and as a result less energy is transferred from the house. Although the rockwool costs more to fit, it saves much more on the energy bills each year and so has a shorter payback time than the sheepwool and so is more cost effective.

Now try this

D-C

1. The U-value for a wall with an air-filled cavity is 1.5. If the cavity is filled with foam the U-value becomes 0.5, reducing the energy lost by two thirds. Explain the change in the U-value when the cavity is filled. *(3 marks)*

B-A*

2. The heating bill for a house is £1200 per year. 35% of energy transfer is through the walls. Calculate the annual saving if the cavities are filled with foam. *(3 marks)*

Specific heat capacity

Different amounts of energy are needed to change the temperature of different materials.

The SPECIFIC HEAT CAPACITY of a material is the amount of energy needed to raise the temperature of 1 kg of the material by 1 °C.

The equation is $E = m \times c \times \theta$

- E is the energy transferred in joules (J)
- m is the mass of the material in kilograms (kg)
- c is the specific heat capacity in J/kg°C
- θ is the temperature change in °C.

Solar panels

Solar panels on roofs use the Sun's energy to heat water. The hot water can be used for washing or to heat the house.

The panels contain pipes. Cold water enters the pipes and is heated up by energy from the Sun. This energy passes from the heated water to cooler water in the tank overnight.

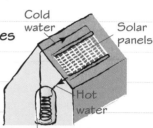
Cold water · Solar panels · Hot water · Hot water tank

Worked example

target **D-C**

1 A solar water heating system heats 80 kg of water from 20 °C to 60 °C.

> Make sure that you are using the correct units for the quantities you use in calculations.

Calculate the amount of energy transferred to the hot water.
The specific heat capacity of water is 4200 J/kg°C.   *(3 marks)*

$E = m \times c \times \theta$

$E = 80\,kg \times 4200\,J/kg°C \times 40°C = 13\,440\,000\,J$

target **B-A***

2 Many radiators that use electrical energy contain an oil. A radiator requires 475 000 J of electrical energy to raise the temperature of 5 kg of oil by 50 °C.

(a) Show that the specific heat capacity of the oil is 1900 J/kg°C.   *(3 marks)*

$c = E/(m \times \theta)$

$c = 470\,000\,J\,/(5\,kg \times 50°C) = 1900\,J/kg°C$

AQA SKILL **Explain** Page 95

(b) Explain why an oil-filled radiator transfers less energy to a room than a similar water-filled radiator at the same temperature (specific heat capacity of water = 4200 J/kg°C).   *(2 marks)*

The oil-filled radiator stores less energy than the water-filled radiator because oil has a lower specific heat capacity than water.

Now try this

target **D-C**

1 Storage heaters contain a ceramic material that is heated by electrical energy at night. The energy is transferred to the room over several hours. The material used may be concrete (specific heat capacity 960 J/kg°C) or brick (900 J/kg°C).

(a) Two storage heaters had their temperatures raised by 60 °C. One contained 100 kg of brick and the other contained 100 kg of concrete. Calculate the energy needed in both cases.   *(4 marks)*

(b) Evaluate the use of concrete and brick in storage heaters.   *(3 marks)*

Energy and efficiency

Energy cannot be created or destroyed, but it can be stored, transferred into useful forms or dissipated. For example, a light bulb transfers electrical energy to light, but some of the energy is dissipated.

Dissipated means the energy spreads out, making the surroundings warmer. It is 'wasted' energy.

Worked example   D-C

AQA SKILL Interpret Page 95

Sankey diagrams show energy transfers.

A kettle is supplied with 200 kJ of electrical energy. 160 kJ of the energy raises the temperature of the water in the kettle.

Sketch a Sankey diagram for the energy transfers in the kettle.   *(3 marks)*

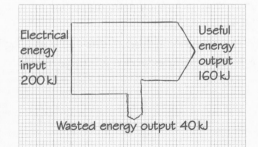

Electrical energy input 200 kJ   Useful energy output 160 kJ

Wasted energy output 40 kJ

Efficiency

$$\text{Efficiency} = \frac{\text{useful energy out}}{\text{total energy in}} \text{ or } \frac{\text{useful power out}}{\text{total power in}}$$

$$\text{Payback time} = \frac{\text{cost of new appliance}}{\text{cost of energy saved each year}}$$

Payback time is the time taken for the cost of a more efficient appliance to be paid by reduced energy bills.

 EXAM ALERT!

The width of the arrows represents the amount of each type of energy. Make sure you can draw a Sankey diagram and understand what a Sankey diagram shows.

Students have struggled with questions like this in recent exams – **be prepared!**

Worked example   B-A*

AQA SKILL Evaluate Page 95

An old type of light bulb uses 80 W of electrical power. It gives out 5 W of light. An LED bulb also gives out 5 W of light but has an efficiency of 0.9. LED bulbs cost about 20 times more than older bulbs but last ten times longer.

(a) Calculate the input energy for the LED.   *(2 marks)*

Energy input = useful energy/efficiency = $\frac{5\,W}{0.9}$ = 5.6 W

(b) Evaluate the cost effectiveness of replacing old style bulbs with LEDs.   *(2 marks)*

The LED uses much less electrical energy than the old bulb, and it does not need to be replaced as often. The payback time for the LED must be less than the life of the LED for it to be cost effective.

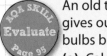 **Now try this**

 target B-A*

1 A new oven costs £450 and has a payback time of 9 years.

   (a) Calculate the saving in energy cost each year.   *(3 marks)*

   (b) Suggest how the new oven costs less to run while delivering the same amount of useful energy.   *(2 marks)*

Physics six mark question 1

There will be one 6 mark question on your exam paper which will be marked for *quality of written communication as well as scientific knowledge*. This means that you need to apply your scientific knowledge, present your answer in a logical and organised way, and make sure that your spelling, grammar and punctuation are as good as you can make them.

Worked example

A coffee shop wants to provide its customers with a cup that will keep a drink hot for longer.

Explain the ways that energy transfer from a cup of hot drink could be reduced. *(6 marks)*

Energy is transferred from an open cup containing a hot drink in various ways.

- Conduction. Energy is transferred from the hot drink into the solid material of the cup and then into the hand holding it. Conduction happens when vibrating particles collide and transfer energy to neighbouring particles. Plastics and paper are poor conductors so should be used for the cup.

- Convection. Energy is transferred to particles of air when they hit a hot surface. Warm air rises and more cold air is drawn in to replace it. Convection can be decreased by trapping the air, such as by fitting a lid.

- Radiation. Hot objects give off more infrared radiation than cold objects. A white or silver cup would emit less radiation than a dark one.

Bullet points
You can use bullet points to structure your answer, but don't just write out a list. The bullet points should contain whole sentences.

As well as the three ways mentioned you could also give evaporation as an example. The faster-moving particles escape from the drink, leaving slower ones behind, so the drink cools. A lid will stop this happening.

Now try this

Use the data in the table to explain the advantages and disadvantages of replacing all the single-glazed windows in a house with double-glazed windows. *(6 marks)*

Component	U-value	Cost in £	Annual reduction in energy bill in £	Payback time in years
Cavity brick wall	1.8			
Single-glazed windows	5.0	2000	0	
Double-glazed windows	2.0	5000	250	20

Electrical appliances

The cost of using electrical appliances can be worked out if we know the amount of energy transferred in a given time.

$$\text{Energy transferred in J} = \text{power in W} \times \text{time in s}$$
$$E = P \times t$$

The cost of electricity supplied by the mains is in pence per kilowatt-hour. Kilowatt-hours can be calculated like this:

Energy transferred in kilowatt-hours = power in kilowatts × time in hours

Take care! **Watts (W)** are a unit of power, but **kilowatt-hours (kWh)** are a unit of energy.

Energy transfers

Electrical appliances carry out many different energy transfers.

Light bulbs, display screens: electricity → light

Kettles, heaters: electricity → energy transfer by heating

Motors: electricity → kinetic energy

Radios: electricity → sound energy

Worked example

target
D-C

1 Electrical energy costs 12p per kilowatt-hour. Use the meters shown on the right to calculate the cost of the electrical energy that was used in 12 hours. *(3 marks)*

 7 4 4 5 (kWh)   9 a.m.

 7 4 6 0 (kWh)   9 p.m.

Electrical energy used = 7460 − 7445
= 15 kWh;

cost = 15 kWh × 12p = 180p or £1.80

2 A tumble dryer uses 1.8 kW of electrical power and runs for 1 hour and 15 minutes. Electrical energy costs 12p per kWh. Calculate the cost of using the tumble dryer during this time. *(3 marks)*

Energy tranferred = 1.8 kW × 1.25 h = 2.25 kWh; cost = 2.25 kWh × 12p = 27p.

Electricity meters measure the amount of electrical energy supplied to all the appliances in a building. The difference in readings taken at two times is the electrical energy used in kWh.

Check that your answers are sensible! The cost of using most appliances is usually less than a pound a day.

Now try this

Remember that 1 watt is 1 joule/second.

target
D-C

1 A battery-powered fan has a power of 2.2 W. Calculate the electrical energy transferred when it is used for 5 minutes. *(2 marks)*

target
B-A*

2 A mobile phone battery is recharged using a mains charger with a power of 0.04 kW. It takes 2 hours to charge up fully. Electricity costs 12p per kilowatt-hour.

(a) Calculate the cost of charging the phone fully. *(2 marks)*

(b) When it is in use the mobile phone transfers electrical energy at a rate of 3.2 W. Calculate the length of time the phone can be used on one full charge. *(3 marks)*

Choosing appliances

Some electrical appliances are more suitable than others for certain applications.

More efficient appliances use less electricity, so they cost less to use. The higher efficiency may be due to stopping energy transfer to or from the appliance by improving the insulation around it. Kettles can be used more efficiently by only using them to heat the amount of water needed.

No mains supply?

In some places, cuts in the electricity supply are common while in others there is no mains electricity supply at all. Electrical energy has to be provided from generators using fossil fuels or a renewable source of energy.

Remember that renewable sources may be unreliable or not available all day, so electrical energy must be stored, most probably in batteries.

Worked example

1 A café owner is considering buying one of the two toasters shown in the table. Evaluate the data to decide which is the most cost effective toaster. *(3 marks)*

Toaster	Electrical power used in kW	Time to toast 100 slices
A	1.5	2 hours
B	2.2	1 hour

Energy transferred in toasting 100 slices: toaster A = 1.5 kW × 2 h = 3 kWh; toaster B = 2.2 kW × 1 h = 2.2 kWh. Thus, although B has the higher power it transfers less energy to do the job and is therefore the most cost effective.

2 An emergency medical centre in a remote area has no access to a mains electricity supply. Its instruments use, on average, 28 kWh of electrical energy each day, supplied by a bank of batteries charged by solar cells and a wind turbine. The centre has a 4 kW solar cell that runs at an average of 70% of capacity for 8 hours a day. It also has a 2 kW wind turbine. The wind blows with sufficient power to charge the batteries for an average of 3 hours a day. Evaluate whether the centre is able to keep its batteries charged. *(3 marks)*

Electrical energy available = 70% × 4 kW × 8 h + 2 kW × 3 h = 28.4 kWh

This is just sufficient to keep the batteries fully charged, but if the sunshine or wind is below average there will be a shortfall in the energy supply. An additional back-up supply is necessary.

Now try this

1 A scientist compared refrigerators used in places where the electricity supply is unreliable. Medicines stored in the fridge must be kept below 6 °C. The graph shows the temperature inside the refrigerators after the power is turned off. Compare the usefulness of the two refrigerators in storing medicines when the electricity supply fails. *(3 marks)*

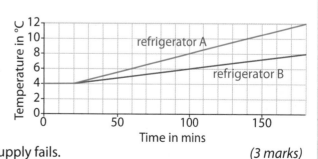

2 The *Whizz* electric bike has a motor with a power of 250 W and a battery that stores 0.52 kWh of electrical energy. A cheaper bike, the *Bikee*, has a power of 200 W and a 0.36 kWh battery. Compare the effectiveness of the two electric bikes. *(4 marks)*

Generating electricity

Most of our electricity is generated using power stations that turn water into steam.

Power station fuels

Energy sources for power stations include:

- FOSSIL FUELS (coal, oil, natural gas)
- NUCLEAR FISSION (the nuclei of uranium or plutonium atoms split up, releasing a lot of energy)
- BIOFUELS (fuels obtained from organisms, such as wood, straw, ethanol, oil palm, gas from rotting waste).

Growing biofuels uses up land that could be used to grow food.

Supplies of some fossil fuels (oil and gas) will only last a few more decades. Fossil fuels, uranium and plutonium are NON-RENEWABLE sources of energy.

Accidents in nuclear power stations are rare but very costly. Nuclear power stations are expensive to build and very expensive to close down (decommission).

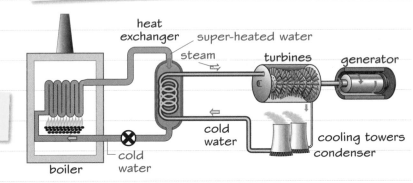

heat exchanger — super-heated water — steam — turbines — generator — cold water — cooling towers — condenser — boiler — cold water

---

Worked example   target B-A*

Compare the use of natural gas and nuclear fuels as sources of energy used to generate electricity. Use the data from the table in your answer.

Source of energy	% of electricity generated in the UK	Cost in p/kWh	Cost to build in £/kW	Lifetime of power station in years	Start-up time	Lifetime of fuel supply in years
natural gas	41	8	700	30	fast	30–50
nuclear	15	9.9	3000	60	slow	100–200

(4 marks)

Gas-fired power stations are cheaper to build and produce electricity most cheaply, but they do have to be replaced more often than nuclear and the gas will run out before nuclear fuel. Gas-fired stations can cope with short peaks in demand because they can be turned on and off easily, while nuclear reactors are slow to start and stop, and so are kept running continuously. Nuclear is thus best for supplying the amount of electricity needed all the time (the base load), while gas is effective for topping up the amount needed at peak times.

Now try this

1 Nuclear power stations have much lower fuel costs than coal-fired stations. Suggest why the cost of electricity from nuclear energy is almost the same as for coal. (2 marks)

2 Many coal and nuclear power stations are nearing the end of their life. Suggest options for coping with this problem in the next ten years.

(4 marks)

Renewables

Renewable sources of energy can also be used to generate electricity.

Renewable sources of energy

Wind: turbines drive electricity generators. They are expensive because a huge number are needed. The wind is unreliable as it does not always blow, but the turbines require little maintenance.

Hydroelectricity: Falling water from reservoirs or rivers turns turbines to run generators. It is a fairly reliable source because rainwater can be stored and used when needed, but dams are very expensive to build.

Tides: Seawater can be trapped behind a barrage at high tide and used to turn turbines. Tidal currents can also be used to turn turbines placed underwater. The tides are predictable and reliable but barrages are very expensive to construct.

Waves: The rise and fall of waves is used to turn turbines. The height of the waves is unpredictable and the machinery needed has not been tested to see how much maintenance is required.

Geothermal: In some volcanic areas hot water and steam from deep underground can be used to drive turbines. The energy is available whenever it is needed, but there are not many sites where it is available.

Solar: Solar cells turn sunlight directly into electricity. The cells are fitted to roofs or set up in fields to form 'solar farms'. Solar cells are expensive and only turn about 20% of the Sun's energy into electricity. They only work in daytime and when the sky is clear, but they need very little maintenance and can be used almost anywhere.

No National Grid?

Solar cells and small wind turbines are used to provide power for road signs. Small-scale hydroelectric generators are used in remote communities. In both cases it is too expensive to connect the generators to the National Grid.

Worked example   target D-C

AQA SKILL
Explain
Page 95

A pumped storage scheme uses spare electrical energy to pump water to a high reservoir when the demand for electricity is low (for example, at night). At other times the water drives hydroelectric generators. Explain why this method is used. *(2 marks)*

Electricity is difficult to store, so instead the electricity is used at quiet times to pump water. When the demand for electricity is high the water can be released to turn turbines and generate electricity for the country.

> The most common way is to store the energy in batteries, which are very expensive.

Now try this

target D-C

target B-A*

1 A mountainous country with a coastline such as Norway needs to replace its fossil fuel power stations. Compare tidal and hydroelectric power as sources of electrical energy. *(4 marks)*

2 An offshore wind farm costs almost twice as much to construct as a modern coal-fired power station with a similar power output, but the overall cost of the electrical energy is only a little higher. Discuss the policy of replacing coal-fired power stations with offshore wind.

(4 marks)

Environment and energy

Generating electricity in various ways has damaging effects on the environment.

You do not need to recall the details of the greenhouse effect and global warming at this stage.

Burning fuels releases pollutants into the atmosphere. Carbon dioxide contributes to global warming and climate change. Other substances cause acid rain.

Coal mines, oil rigs, power stations and wind farms are thought by some people to be visual pollution.

Noise pollution may be a disadvantage of wind farms. Mines are also noisy.

Environmental effects of using energy resources

Habitats can be destroyed by mining and oil and gas collection. Farming biofuels destroys natural habitats. Hydroelectric schemes and tidal barrages can cause habitats to be damaged.

Burning solid fuels (coal, wood) produces ash, which must be disposed of.

Burning biofuels does not contribute as much as fossil fuels to global warming because the amount of carbon dioxide released is the same as was absorbed when the plants were growing.

Nuclear power stations take up a lot less space than stations that burn coal because they do not need space for storing fuel. Wind farms cover many square kilometres to produce the same amount of power as a small power station.

Worked example D–C

It has been suggested that empty North Sea oil and gas fields could be used as places to store carbon dioxide produced by power stations burning fossil fuels.

1 Explain why carbon capture and storage (CCS) is necessary. *(2 marks)*

Carbon dioxide is produced when fossil fuels are burned. Carbon dioxide levels are rising in the atmosphere and are linked to global warming.

CCS is expensive and suitable sites have not been tested yet.

2 State **one** reason why CCS has not been carried out yet. *(1 mark)*

The technology to collect and store carbon dioxide underground is still being developed.

3 Why are the North Sea oil and gas field a suitable storage places? *(2 marks)*

As they held oil and gas for millions of years it is thought that the carbon dioxide could be stored and wouldn't be released into the atmosphere.

Now try this

1 Using carbon capture and storage (CCS) doubles the cost of coal-fired power stations, making it more expensive than natural gas, nuclear and some renewable sources of energy. Suggest **one** advantage and **one** disadvantage of using coal-fired power stations with CCS to generate electricity. *(2 marks)*

2 Anaerobic digesters convert food waste to methane gas, which is burned to generate electricity. If it is left in landfill sites the waste would form methane naturally. Methane is a powerful greenhouse gas. Suggest **two** environmental benefits of constructing anaerobic digesters. *(2 marks)*

Distributing electricity

In the UK, the National Grid distributes electricity from power stations to consumers.

The National Grid

Mains electricity is distributed to homes, shops, offices and factories by cables from power stations, which may be hundreds of kilometres away.

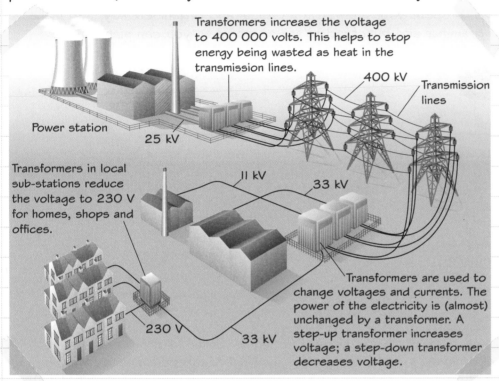

Transformers increase the voltage to 400 000 volts. This helps to stop energy being wasted as heat in the transmission lines.

400 kV

Transmission lines

Power station

25 kV

Transformers in local sub-stations reduce the voltage to 230 V for homes, shops and offices.

11 kV

33 kV

230 V

33 kV

Transformers are used to change voltages and currents. The power of the electricity is (almost) unchanged by a transformer. A step-up transformer increases voltage; a step-down transformer decreases voltage.

When electrical energy travels along a wire some of it is dissipated. Increasing the voltage to 400 kV reduces current, which reduces the amount of energy wasted.

High voltage is very dangerous. The electric shock caused by touching a high-voltage cable can kill.

Worked example   target D-C

AQA SKILL Compare Page 95

Compare the advantages and disadvantages of burying high-voltage cables underground with hanging them overhead from pylons.

(4 marks)

Advantages: Pylons are a form of visual pollution. People can be injured if they climb pylons but are unlikely to be injured by buried cables.

Note where the step-up and step-down transformers are in the National Grid. You do not need to know how transformers are built or work but you do need to know what they do and why.

Disadvantages: Burying cables is much more expensive than carrying them on pylons. It is more difficult to make repairs to buried cables than to those above ground. Digging the trenches damages habitats temporarily.

Now try this

target D-C

target B-A*

1 Describe how step-up and step-down transformers are used in the National Grid.   *(2 marks)*

2 Campaigners in an area of outstanding natural beauty are pressing the National Grid to install a new 400 kV power line underground instead of above ground. Burying power lines costs about £20 million per km while hanging the cables from pylons costs about £2 million per km.

(a) State the advantages of distributing electrical energy at 400 kV.   *(2 marks)*

(b) Discuss the reasons why the National Grid should or should not agree to the campaigners' demands.   *(2 marks)*

Physics six mark question 2

There will be one 6 mark question on your exam paper which will be marked for *quality of written communication as well as scientific knowledge*. This means that you need to apply your scientific knowledge, present your answer in a logical and organised way, and make sure that your spelling, grammar and punctuation are as good as you can make them.

Worked example

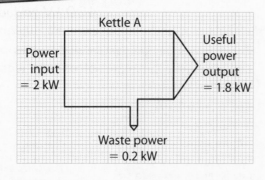
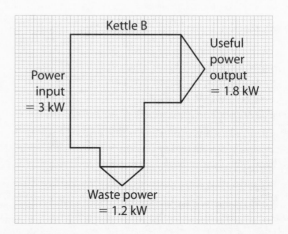

A scientist tested two kettles to see which was the most efficient. He boiled 1 litre of water in each kettle and the water took 3 minutes to boil. Use the information in the Sankey diagram to compare the efficiency of the two kettles. *(6 marks)*

The two kettles provide the same amount of useful power (1.8 kW), although kettle B uses more power (3 kW instead of 2 kW). This is because kettle A is more efficient.

This means that less of the energy provided to kettle A is wasted than for kettle B. The wasted energy is transferred as heat energy, increasing the temperature of the kettle and the surrounding air.

The efficiency of kettle A is 1.8 kW/2 kW = 0.9

The efficiency of kettle B is 1.8 kW/3 kW = 0.6

Although both kettles take the same length of time to boil 1 litre of water, kettle A uses less energy.

> You should explain terms such as **efficiency**.

> The question says to **compare** the two kettles, so you should write about both kettles, saying what is similar or different about them.
>
> You should use data from the diagrams and from calculations to work out a full answer and be able to compare the kettles.

Now try this

Burning fossil fuels provided most of the electricity generated in the UK in 2011. Only a small percentage was provided by biofuels, hydroelectric, wind and solar power, which are renewable sources of energy. Describe the impact of fossil fuels and renewable sources of energy on the environment. *(6 marks)*

Properties of waves

WAVES transfer ENERGY from one place to another. Waves are carried by something that is oscillating.

Transverse VS ## Longitudinal

water and electromagnetic waves move like this

transverse wave

direction of wave travel
direction of energy transfer

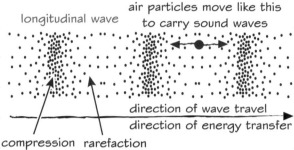
air particles move like this to carry sound waves

longitudinal wave

direction of wave travel
direction of energy transfer

compression   rarefaction

Mechanical waves must move through solids, liquids or gases. They cannot travel through a vacuum. They may be transverse or longitudinal.

EXAM ALERT!

Make sure you can describe the difference between a transverse wave and a longitudinal wave.

Students have struggled with this topic in recent exams – **be prepared!**

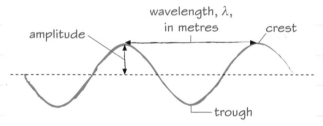
amplitude
wavelength, λ, in metres   crest
trough

Frequency, *f*, is the number of complete waves passing a point in 1 second, in hertz, Hz. The wave equation is, $v = f \times \lambda$, where *v* is the velocity in metres per second, m/s

---

Worked example

 target **D–C**

1 The sound of a whistle has a frequency of 680 Hz and a wavelength of 0.5 m.
Calculate the speed of the sound through air.
(2 marks)

$$v = f \times \lambda = 680 \text{ Hz} \times 0.5 \text{ m}$$
$$= 340 \text{ m/s}$$

Make sure the data you have been given are in the correct units.

 target **B–A***

2 The speed of light in a vacuum is 3×10^8 m/s.
Calculate the wavelength of red light that has a frequency of 4×10^{14} Hz.

$$\lambda = v/f = \frac{3 \times 10^8 \text{ m/s}}{4 \times 10^{14} \text{ Hz}}$$
$$= 7.5 \times 10^{-7} \text{ m (or 0.75 micrometres)}$$

Pick out the data you have been given and find the equation you need on the equations sheet.

Write the equation out and show how you get your answer.

Now try this

 target **B–A***

1 A tap drips twice a second. The drops fall into a bowl of water, producing ripples. There are four ripples between where the drips enter the water and the side of the bowl 16 cm away.
(a) Explain why only energy is carried by the waves to the side of the bowl.   *(2 marks)*
(b) Calculate the speed of the waves across the water in the bowl.   *(4 marks)*

Electromagnetic waves

ELECTROMAGNETIC waves are transverse waves that include light, radio waves and X-rays. All of these waves travel at the same speed in a vacuum. The different types of electromagnetic wave form a continuous spectrum.

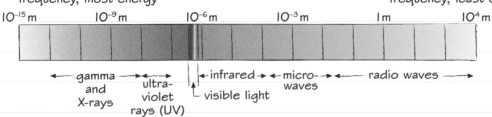

shortest wavelength, highest frequency, most energy

longest wavelength, lowest frequency, least energy

10^{-15} m 10^{-9} m 10^{-6} m 10^{-3} m   1 m 10^4 m

gamma and X-rays   ultra-violet rays (UV)   ←infrared→ ←micro-waves→ ←──── radio waves ────→

visible light

'Continuous' means that there are no breaks in the spectrum and one type of wave merges into the next.

Hazards of electromagnetic waves:
- Gamma and X rays cause cancer.
- UV causes sunburn and skin cancer.
- Visual can damage eyes.
- Infrared can cause burning.

Communications

Some electromagnetic waves are used in communication because the waves can carry information with the energy they transfer.

Electromagnetic wave	Use
radio	television and radio
microwave	mobile phones and satellite TV
infrared	remote controls
visible	photography

 Worked example target D-C

AQA SKILL Compare Page 95

Compare the electromagnetic waves used for short-range communications (e.g. remote controls) and long-range communications (e.g. satellite TV).

(4 marks)

Infrared radiation is used in remote controls. It has a higher frequency and shorter wavelength than microwave radiation used in satellite TV. Both types of wave travel at the same speed in a vacuum.

 Worked example target B-A*

AQA SKILL Evaluate Page 95

Between 1992 and 2008, the percentage of the population with mobile phones went from nearly 0% to almost 100%. The number of brain tumours remained at the same level that it had been at before.

A scientist concluded that there was no link between mobile phone and brain tumours. Evaluate the scientist's conclusion.

(2 marks)

As there was no change in the number of brain tumours in the period the conclusion would seem justified unless the tumours caused by phone use take a long time to appear, or are rare.

Make sure that you use the data you have been given in the question. This is just one study. Some other studies suggest there is a link between mobile phones and cancer.

Now try this

 target B-A*

1 A mobile phone company said that none of the 100 children given a free phone were harmed by the microwave radiation. Evaluate this study.

(3 marks)

Waves

 1 REFLECTION happens when a wave bounces off the interface between two materials. The wave changes direction but doesn't cross the interface.

 2 REFRACTION happens when a wave crosses an interface between two materials. The wave changes direction unless the rays hit the interface at right angles.

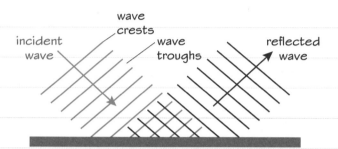

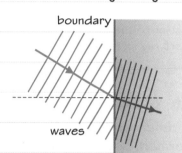

 3 DIFFRACTION happens when waves pass through a gap or over an obstacle that is similar in size to the wavelength of the wave. The wave spreads out through the gap.

A ray is a line showing the direction the waves are travelling.

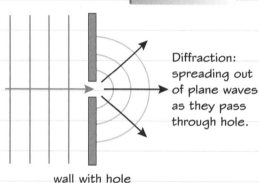

Diffraction: spreading out of plane waves as they pass through hole.

wall with hole

Worked example   target D-C

 IQA SKILL Explain Page 95

Explain why it is possible to pick up radio signals in a valley but not a TV or mobile phone signal. *(3 marks)*

Radio, TV and mobile phone signals can pass through the walls of buildings.

Radio waves have a wavelength similar to the height of hills, so they are diffracted around the hills. This means that radio waves reach places that cannot see the transmitter. Shorter wavelengths are used for TV and mobile phones, which are not diffracted by hills so there must not be any hills between your TV or mobile phone and the transmitter.

Now try this

 target D-C

1 Compare what happens to waves when they are reflected and refracted at a boundary between two materials. *(2 marks)*

 target B-A*

2 Suggest explanations for the following observations. You can sketch diagrams to help your explanations.
  (a) An infrared remote control will sometimes operate a TV when pointed at a wall. *(2 marks)*

(b) A straw in a glass of water looks bent. *(2 marks)*

(c) When standing outside a room with the door open you can hear someone in the room talking even though you cannot see them. *(2 marks)*

(d) When an X-ray beam is directed at a crystal the rays are diffracted, forming a distinctive pattern. This does not happen with visible light. *(3 marks)*

Reflection in mirrors

When waves are reflected, the angle of incidence is equal to the angle of reflection.

The INCIDENT RAY shows the direction of waves moving towards a surface.

The NORMAL is an imaginary line drawn perpendicular to the surface where the incident ray meets it.

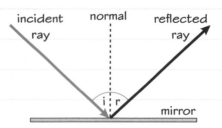

The REFLECTED RAY shows the direction after reflection.

Images in plane mirrors

When light is reflected from a PLANE mirror, an image is formed, which is VIRTUAL.

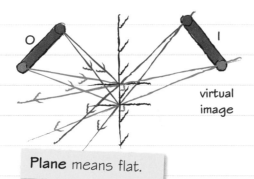

virtual image

Plane means flat.

Virtual means that the rays do not actually pass through the image. Your image in a mirror appears to be behind the mirror but the light does not come from there.

Worked example   D-C

A man uses a mirror to help him shave. Draw the incident and reflected rays on the diagram to show how he can see his chin. *(2 marks)*

Don't forget to add arrows when drawing rays to show the direction the light is travelling. The light travels towards the eye.

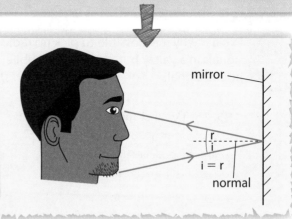

mirror

i = r

normal

Now try this

D-C **1** A driveway reaches the road at a right angle. A driver on the driveway uses a mirror on the opposite side of the road to see cars approaching from the left. Draw a labelled ray diagram showing how the driver can see a car coming from the left while looking straight ahead.

(3 marks)

B-A* **2** A hairdresser holds up a mirror behind you so that you can see the back of your head in the mirror in front of you. Draw a diagram showing how you see a point on the back of your head.

(3 marks)

Sound

Sound waves are longitudinal waves that cause vibrations in materials.

The material that sound waves travel through is known as the MEDIUM. Sound waves can travel through solids, liquids and gases. The PITCH of a sound is determined by its frequency. A high-pitched sound is a wave with a high frequency. The LOUDNESS of a sound is related to the amplitude of the vibrations.

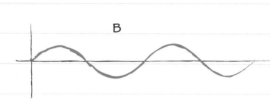

> It is easier to show sound waves as if they are displayed on an oscilloscope. The higher the peak, the more the particles are compressed.

A and B are two sound waves. B has a lower pitch than A but A is louder.

Echoes

Sound waves can be reflected off surfaces. We hear the reflections as ECHOES. Echoes arrive after the main sound because the waves have travelled further.

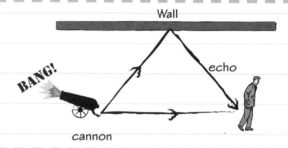

Worked example   **B-A*** target

1 You are listening to the radio.

(a) Describe how the sound reaches your ear from the radio. *(2 marks)*

As the sound wave passes through the air the particles of air vibrate. Sometimes they become compressed and at other times they spread out or become rarefied. This is a longitudinal wave. The air particles do not travel from the radio to the ear.

(b) Explain what happens if the sound becomes louder. *(2 marks)*

The amplitude of the sound waves increases so the amount of compression and rarefaction increases.

(c) Explain what happens when the sound rises in pitch. *(2 marks)*

The frequency increases. The zones of compression are closer together so that more waves arrive at the ear each second.

EXAM ALERT!

Remember that the pitch of a sound is determined by its frequency and that the loudness of a sound is determined by its amplitude.

> Students have struggled with exam questions similar to this – **be prepared!**

Now try this

 target **D-C**

1 Explain how whales communicate with each other even when they are a large distance apart under water. *(2 marks)*

2 Explain why an echo always arrives after the original sound. *(2 marks)*

 target **B-A***

3 Look at the diagrams of sound waves shown at the top of the page. Sketch a diagram of a wave that has a pitch and loudness greater than both wave A and wave B. *(2 marks)*

Red-shift

The Doppler effect can be used to measure the speed and direction of moving objects.

The Doppler effect

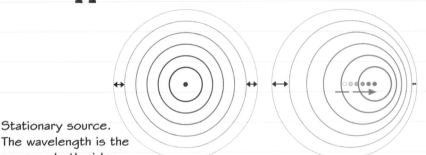

Stationary source.
The wavelength is the
same on both sides.

When a source of a wave is moving, the waves in front of it get squashed. Their frequency increases and wavelength decreases. The waves behind the source become stretched. Their frequency is lower and wavelength longer.

The faster the source moves, the bigger the change in frequency and wavelength. If the source of a sound is coming towards us we hear a higher-pitched sound. If the source is moving away the sound is at a lower pitch.

Red-shift

The radiation from stars in distant galaxies is at longer wavelengths than radiation from similar stars in our own galaxy. This is called red-shift. Red-shift shows that the galaxies are moving away from us.

The further the galaxies are from us the bigger the red-shift. This means that the further away the stars are the faster they are moving away from us.

galaxy moving away

The Doppler effect does not just apply to visible light, but to radio, microwave and other electromagnetic waves too.

Worked example

target D-C

1 An astronomer observes two galaxies. The light from galaxy A is red-shifted more than galaxy B. State **two** conclusions the astronomer can make. *(2 marks)*

Galaxy A is moving away from Earth faster than B. Galaxy A is further away than B.

target B-A*

2 A blind person is a witness at an incident involving two identical fire engines. Both had their sirens sounding. The witness reports that one of the fire engines stopped while the other continued to approach. Explain how the witness could be correct without being able to see what happened. *(2 marks)*

The witness would have heard each siren at a different pitch because the frequency of the approaching fire engine would be higher than the one that had stopped due to the Doppler effect.

Now try this

target D-C

1 Describe the observations that reveal to astronomers that distant galaxies are moving away from us faster than galaxies that are closer. *(2 marks)*

target B-A*

2 Police use speed guns to determine the speed of cars. The gun emits a microwave beam at a fixed frequency and it contains a microwave receiver. Suggest how the gun can be used to catch speeding motorists. *(3 marks)*

The expanding universe

Observations of red-shift provide evidence for theories about the universe.

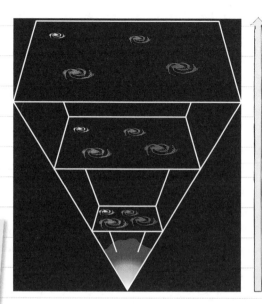

Red-shift shows astronomers that the universe is expanding. The Big Bang theory suggests that the universe began from a tiny point and expanded to become what we see now.

Time

The Big Bang theory predicts that the radiation released when the universe started to expand is still present. As the universe expanded the wavelength increased so that it is now microwaves. This cosmic microwave background radiation (CMBR) has been observed and is evidence for the Big Bang theory.

The Big Bang is the only theory for the origin of the universe that is accepted by scientists today.

The CMBR seems to come from every part of space as predicted by the Big Bang theory.

The Big Bang is not like a normal explosion, which blows material into space. Since the Big Bang, space itself has expanded so that everything in the universe has moved apart.

Worked example B-A*

AQA SKILL
Explain
Page 95

1 There have been many theories for why the universe is as it is today. Explain why only the Big Bang theory is accepted by almost all scientists.   *(2 marks)*

The Big Bang theory is the only theory that explains the evidence, including the red-shift of distant galaxies, and it predicted the presence of the CMBR in every part of space.

2 The Big Bang theory successfully explains observations made by astronomers but has limitations, like all scientific models. State **two** limitations of the Big Bang theory.   *(2 marks)*

The model cannot describe what happened before the Big Bang or what happens beyond the furthest part of the universe we can see.

The Big Bang also cannot explain why the expansion of the universe is getting faster.

Now try this

target
D-C

1 The Big Bang theory is accepted by most scientists because it explains the observations that astronomers have made. Give **two** pieces of evidence for the Big Bang theory.   *(4 marks)*

target
B-A*

2 Explain why the Big Bang theory is accepted by most scientists.   *(2 marks)*

Physics six mark question 3

There will be one 6 mark question on your exam paper which will be marked for *quality of written communication as well as scientific knowledge*. This means that you need to apply your scientific knowledge, present your answer in a logical and organised way, and make sure that your spelling, grammar and punctuation are as good as you can make them.

Worked example

Compare the properties and uses of microwaves and radio waves. *(6 marks)*

Microwaves and radio waves are both types of electromagnetic radiation. They are transverse waves that can travel through a vacuum at the speed of light. Radio waves have the longer wavelength and therefore have the lower frequency. Microwaves carry more energy. They can be used in microwave cookers to transfer energy to food.

Both types of radiation can be reflected by surfaces, absorbed or refracted when they pass through materials and diffracted when they pass through gaps.

Both types of radiation are used in communications. The longer wavelength of radio waves means that they can be diffracted around buildings and hills. This means that radio receivers can pick up radio communications even if they are not in line of sight with the transmitter. The shorter wavelength of microwaves means that a mobile phone must be on a straight line from a transmitter.

Writing your answer

The question uses the command COMPARE, so you should describe the similarities *and* differences in microwave and radio waves, and their uses in communications.

Write your answers in full sentences and check your spellings, particularly of technical words.

 Recall the properties that all types of electromagnetic radiation have.

 Show that you understand that the wavelength and frequency of waves are related.

 Remember that all waves can be reflected, refracted and diffracted.

Now try this

Most scientists agree that the Big Bang theory explains how the universe we see today was formed. Describe what the Big Bang theory is and the evidence that supports it. *(6 marks)*

Periodic Table

Key:
- relative atomic mass
- **atomic symbol**
- name
- atomic (proton) number

Example: 1 **H** hydrogen 1

Group 1	Group 2												Group 3	Group 4	Group 5	Group 6	Group 7	Group 0
																		4 **He** helium 2
7 **Li** lithium 3	9 **Be** beryllium 4												11 **B** boron 5	12 **C** carbon 6	14 **N** nitrogen 7	16 **O** oxygen 8	19 **F** fluorine 9	20 **Ne** neon 10
23 **Na** sodium 11	24 **Mg** magnesium 12												27 **Al** aluminium 13	28 **Si** silicon 14	31 **P** phosphorus 15	32 **S** sulfur 16	35.5 **Cl** chlorine 17	40 **Ar** argon 18
39 **K** potassium 19	40 **Ca** calcium 20	45 **Sc** scandium 21	48 **Ti** titanium 22	51 **V** vanadium 23	52 **Cr** chromium 24	55 **Mn** manganese 25	56 **Fe** iron 26	59 **Co** cobalt 27	59 **Ni** nickel 28	63.5 **Cu** copper 29	65 **Zn** zinc 30		70 **Ga** gallium 31	73 **Ge** germanium 32	75 **As** arsenic 33	79 **Se** selenium 34	80 **Br** bromine 35	84 **Kr** krypton 36
85 **Rb** rubidium 37	88 **Sr** strontium 38	89 **Y** yttrium 39	91 **Zr** zirconium 40	93 **Nb** niobium 41	96 **Mo** molybdenum 42	99 **Tc** technetium 43	101 **Ru** ruthenium 44	103 **Rh** rhodium 45	106 **Pd** palladium 46	108 **Ag** silver 47	112 **Cd** cadmium 48		115 **In** indium 49	119 **Sn** tin 50	122 **Sb** antimony 51	128 **Te** tellurium 52	127 **I** iodine 53	131 **Xe** xenon 54
133 **Cs** caesium 55	137 **Ba** barium 56	139 **La** lanthanum 57	178 **Hf** hafnium 72	181 **Ta** tantalum 73	184 **W** tungsten 74	186 **Re** rhenium 75	190 **Os** osmium 76	192 **Ir** iridium 77	195 **Pt** platinum 78	197 **Au** gold 79	201 **Hg** mercury 80		204 **Tl** thallium 81	207 **Pb** lead 82	209 **Bi** bismuth 83	210 **Po** polonium 84	211 **At** astatine 85	222 **Rn** radon 86
223 **Fr** francium 87	226 **Ra** radium 88	227 **Ac** actinium 89	261 **Rf** rutherfordium 104	262 **Db** dubnium 105	266 **Sg** seaborgium 106	264 **Bh** bohrium 107	277 **Hs** hassium 108	268 **Mt** meitnerium 109	271 **Ds** darmstadtium 110	272 **Rg** roentgenium 111								

The lanthanides (atomic numbers 58–71) and the actinides (atomic numbers 90–103) have been omitted.

Elements with atomic numbers 112–118 have been reported but not fully authenticated.

Cu and Cl have not been rounded to the nearest whole number.

Chemistry Data Sheet

Reactivity series of metals

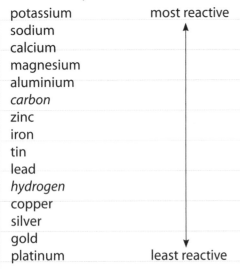

potassium   most reactive
sodium
calcium
magnesium
aluminium
carbon
zinc
iron
tin
lead
hydrogen
copper
silver
gold
platinum   least reactive

Elements in italics, though non-metals, have been included for comparison.

Formulae of some common ions

Positive ions

Name	Formula	Name	Formula
hydrogen	H^+	copper(II)	Cu^{2+}
sodium	Na^+	magnesium	Mg^{2+}
silver	Ag^+	zinc	Zn^{2+}
potassium	K^+	lead	Pb^{2+}
lithium	Li^+	iron(II)	Fe^{2+}
ammonium	NH_4^+	iron(III)	Fe^{3+}
barium	Ba^{2+}	aluminium	Al^{3+}
calcium	Ca^{2+}		

Negative ions

Name	Formula	Name	Formula
chloride	Cl^-	nitrate	NO_3^-
bromide	Br^-	oxide	O^{2-}
fluoride	F^-	sulfide	S^{2-}
iodide	I^-	sulfate	SO_4^{2-}
hydroxide	OH^-	carbonate	CO_3^{2-}

Physics Equations Sheet

$E = m \times c \times \theta$	E energy transferred m mass θ temperature change c specific heat capacity
$\text{efficiency} = \dfrac{\text{useful energy out}}{\text{total energy in}} \ (\times 100\%)$	
$\text{efficiency} = \dfrac{\text{useful power out}}{\text{total power in}} \ (\times 100\%)$	
$E \times P \times t$	E energy transferred P power t time
$v = f \times \lambda$	v speed f frequency λ wavelength

AQA specification skills

In your AQA exam there are certain **skills** that you sometimes need to **apply** when answering a question. Questions often contain a particular **command word** that lets you know this. On this page we explain how to spot a command word and how to apply the required skill.

Note: Watch out for our Spec Skills sticker – this points out the Worked Examples that are particularly focussed on applying skills.

Command word	Skill you are being asked to apply
AQA SKILL Analyse Page 95	**Analyse** the data you are given to answer the question. The data might be a table or a graph. Make sure you refer to the data in your answer. For example: 'Y is twice as strong as X.'
AQA SKILL Compare Page 95	**Compare** how two things are similar or different. Make sure you include both of the things you are being asked to compare. For example: 'A is bigger than B, but B is lighter than A.'
AQA SKILL Consider Page 95	You will be given some information and you will be asked to **consider** all the factors that might influence a decision. For example: 'When buying a new fridge the family would need to consider the following things …'
AQA SKILL Describe Page 95	**Describe** a process or why something happens in an accurate way. For example: 'When coal is burned the heat energy is used to turn water into steam. The steam is then used to turn a turbine …'
AQA SKILL Evaluate Page 95	This is the most important one! Many of the skill statements start with **evaluate**. You will be given information and will be expected to use that information plus anything you know from studying the material in the specification to look at evidence and come to a **conclusion**. For example, if you were asked to evaluate which of two slimming programmes was better, then you might comment like this: 'In programme A people lost weight quickly to start with but then put the weight back on by the end of the sixth month. In programme B they did not lose weight so quickly to start with, but the weight loss was slow and steady and no weight was gained back by the end of the year. I therefore think that programme B is most effective.'
AQA SKILL Explain Page 95	State what is happening and **explain** why it is happening. If a question asks you to explain then it is a good idea to try to use the word 'because' in your answer. For example: 'pH 2 is the optimal pH for enzymes in the stomach because the stomach is very acidic.'
AQA SKILL Interpret Page 95	**Interpret** the data given to you on graphs, diagrams or in tables to help answer the question. Sometimes this kind of question also asks you to **draw** or sketch something. For example: 'Look at the data given and sketch a pyramid of biomass.'
AQA SKILL Identify Page 95	You might be asked to **identify** features of something. For example: 'A polar bear has a thick coat. This prevents the loss of heat energy to the cold environment.'
AQA SKILL Suggest Page 95	You will be given some information about an unfamiliar situation and asked to **suggest** an answer to a question. You will not have learned the answer – you need to **apply** your knowledge to that new situation. For example: 'I think that blue is better than green because …' or 'It might be because …'

ISA support

25% of each Science GCSE comes from controlled assessments called ISAs (Investigative Skills Assignments). Your teacher will explain to you what you have to do but these pages help you with the two papers which are part of every ISA.

All the examples on this page and the next are from Paper 1. The examples are taken from an investigation into measuring the energy released when a piece of bread is burned and how changing the mass of the bread changed the energy released.

Worked example

Research sources (Paper 1)

1 Name two sources you used for your research. Which sources did you find most useful and why?

My sources were:

AQA GCSE Biology (Editor = Nigel English, publisher = Longman)

http://www.nuffieldfoundation.org/practical-biology/how-much-energy-there-food

Source 1 was useful because it explained the scientific terms biomass and how this relates to energy. Source 2 was useful because it showed me a simple method I could use in school.

> This is a very good answer as it gives the title, author and publisher of the book and full URL of the website. It also gives a reason for why each source was useful.

2 In this investigation, you will need to control some of the variables.
   Write down **one** variable that will need to be controlled.

I will need to control the volume of the water.

> In this experiment you could also have said that you needed to control the distance between the bread and the water or the type of bread used. Think about the things you need to keep the same each time. This is what you might have called 'fair testing'.

Preliminary experiments (Paper 1)

3 Describe briefly how you would carry out a preliminary investigation to find a suitable value to use for this control variable.
   You should also explain how the results of this work will help you to decide on the best value for this variable.

I will burn a small sized piece of bread under a boiling tube of water. I will measure how much the temperature rises if I use a small volume of water (10 cm^3), a medium volume of water (25 cm^3) and a large volume of water (50 cm^3). As I am using a small piece of bread, this will be my smallest temperature rise and I need it to be somewhere around 10–20°C rise. I will use the volume in my investigation that gives a temperature rise in this range so that it is clearly measurable, but not so big that a larger piece of bread would heat the water up to its boiling point.

> This is a good answer because it describes in detail how the preliminary work will be done and what you are specifically looking for. It discusses the idea of looking for values that give measurable results across the range of values of the independent variable (the amount of bread burned).

ISA support

Worked example

Plan (Paper 1)

Describe how you plan to do your investigation to test the hypothesis given. You should include:

- the equipment you plan to use
- how you will use the equipment
- the measurements you are going to make
- how you will make it a fair test
- a risk assessment

Apparatus

Dried bread, balance, boiling tube, stand, clamp, boss, metal bottle top, measuring cylinder, thermometer, heat proof mat, Bunsen burner

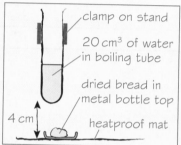

clamp on stand

$20\,cm^3$ of water in boiling tube

dried bread in metal bottle top

4 cm

heatproof mat

Method

1. I will collect some dried bread and cut it into different five sized pieces.

2. I will find the mass of each piece on a balance with a resolution of at least 0.1 g.

3. I will place $20\,cm^3$ of water in a boiling tube using a measuring cylinder.

4. I will clamp the boiling tube 4 cm above the bench.

This is not the full method for this experiment but it shows a good way of setting out your plan.

Risk assessment:

The main dangers in this experiment are from the Bunsen burner flame, burning bread and hot apparatus. I will use tongs to move the hot bottle top, I will handle the heated boiling tube by the top and not the bottom, and I will keep the Bunsen burner towards the back of the bench on the yellow safety flame when not in use. I will also wear eye protection throughout the experiment.

Table (Paper 1)

Draw a table for your results.

Mass of bread in grams	Temperature in °C		
	Start	End	Rise

Variables

Independent variable = mass of bread

Dependent variable = temperature rise

Control variables = volume of water, distance between bread and boiling tube

You don't need to write the variables out like this but sometimes it helps you to be clear what you are controlling, what you are changing and what you are measuring.

It is a good idea to list all the apparatus that you use and to label the diagram.

Your instructions need to be clear. A good guide is to write the instructions so that someone else in your class could follow them to do your experiment. They need to be written in good English, using full sentences, but can be numbered. A diagram can be helpful.

You need to identify real risks and suggest ways to prevent these problems happening.

This is a very good table as it gives columns for the independent variable (mass of bread) and the dependent variable (temperature rise). It also gives the units. The table also includes columns to record the start and end temperatures from which the temperature rise is calculated.

ISA support

Paper 2 of the ISA looks at your own results and those from other similar case studies. In many of the questions, it is important to remember to use the data to justify your answers.

The next few examples relate to an experiment where a student investigated how changing the temperature of an oil affects the viscosity of the oil. They did this by timing how long it took for some vegetable oil to flow through the hole in the bottom of a yoghurt pot. Here are their results.

Temperature in °C	Time for oil to flow through the hole in seconds
21	89
32	46
39	35
53	26
62	18

Worked example

Graphs (Paper 2)

You will be awarded up to 4 marks for your chart or graph.

Here is the graph of my results.

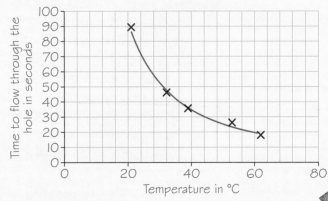

Conclusions (Paper 2)

The hypothesis: *Changing temperature affects the viscosity of oils.*

Do the results support the hypothesis you were given?

Explain your answer.

My results do support my hypothesis. As the temperature increased, the time taken for the oil to flow through the hole decreases and so the oil is less viscous. For example, at 21 °C the oil takes 89 s to flow through the hole, but at 62 °C it takes 18 s.

You will always have to plot a bar chart or line graph. Plot a bar chart for categoric variables, that is those that have words, e.g. colours. Plot a line graph for continuous variables, that is those that are numbers, e.g. temperature. The independent variable should be on the horizontal axis with the dependent variable on the vertical axis.

This is a good graph. The axes have suitable scales. Both axes are labeled and have units. A good best fit line has been plotted that is a smooth and has a similar number of points above and below the line. Best fit lines can be straight or curved.

This is good because it both gives a clear conclusion but also quotes data from the results to support the conclusion.

ISA support

Worked example

Here is the data from three Case Studies that were carried out.

Case study 1

A student timed how long it took for a drop of vegetable oil to flow down a sloping tile at different temperatures.

Temperature of tile and oil in °C	Time for oil to flow down the tile in seconds
20	56
40	24
60	18
80	11
100	8

Case study 2

A student timed how long it took for different vegetable oils to flow through a funnel into a bottle.

Type of vegetable oil	Time for oil to flow through funnel into bottle in seconds
Sunflower oil	45
Rapeseed oil	56
Olive oil	49
Grapenut oil	38

Case study 3

A motor oil company tested the viscosity of its car engine oil. They timed how long it took 20 cm³ of engine oil to flow through a metal tube that was held at an angle.

Temperature of engine oil in °C	Time for oil to flow down tube in seconds
50	125
100	73
150	41
200	27

Draw a **sketch graph** of the results in **Case Study 3**.

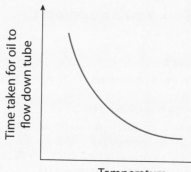

Using case studies (Paper 2)

Explain whether or not the results on the Secondary Data sheet support the hypothesis you were given.

To gain full marks your explanation should include appropriate examples from the results in Case Studies 1 and 2.

Case study 1 supports my hypothesis. As the temperature increases from 20°C to 100°C, the time it takes to flow down the tile falls from 56s to 8s. Case study 2 is irrelevant as the type of oil was changed rather than the temperature.

This answer is good because it clearly states whether both of the case studies support the hypothesis or not, and quotes data to justify the answer.

This is a good sketch graph because both axes are labeled and the line is the correct shape. As this is only a sketch graph to indicate the shape there is no need to have units on the labels. There is also no need to put numbers on the axes but it might help you to draw your sketch to do so.

Answers

You will find some advice next to some of the answers. This is written in italics. It is not part of the mark scheme but just gives you a little more information.

Biology answers

1. A healthy diet

1. Not getting the right balance of energy and nutrients **(1)**.
2. (a) The 19–50 group because their recommended intake is the highest and so they need the most iron **(1)**.
   (b) The developing baby will also need iron **(1)**; so a pregnant woman needs more iron than other women and may not get enough extra iron in her diet **(1)**.

2. Controlling mass

1. Eating less means that you take in less energy **(1)**; exercising more increases the amount of energy you expend **(1)**. If you can expend more energy than you take in, you will lose mass **(1)**.
2. The chart shows that all slimming programmes caused a greater weight loss than exercise only, both after 12 weeks and after a year **(1)**. Of the slimming programmes, programme A and programme C caused similar weight loss after 12 weeks **(1)**, but programme A gave the best weight loss after a year, so it is the best weight loss programme **(1)**.

3. Lifestyle and disease

1. Exercise can help to control weight, and so reduce obesity **(1)**, which is linked to many health problems such as Type 2 diabetes **(1)**.
2. The study is a very large study and so the results are valid **(1)**. The study is only of American men, so the conclusion may not apply to other groups of people **(1)**. The conclusion does not distinguish between fit and unfit men who are not overweight, and these two groups may have differences in their risk factors for health that also cause some of the differences between fit and unfit overweight people **(1)**.

4. Pathogens and infection

1. Washing cleans pathogens off hands **(1)**. This prevents the pathogens from one patient infecting the next patient that the doctor examines **(1)**.
2. When we are infected, it takes a while for pathogens to reproduce inside us **(1)** and make a large enough number of pathogens to make us feel ill **(1)**.
3. Any two suitable reasons that clearly show that the number of microorganisms decreased in a gradual way, or that methods for reducing transmission between patients were developed on a gradual basis. For example: hand washing was introduced to some wards in 2007/08 **(1)**, and then to other wards over the following years **(1)**; OR hand washing started in 2007/08 **(1)**, but more effective cleaning solutions were used over the following years **(1)**.

5. The immune system

1. The antibodies that destroy the pathogen that causes measles will not destroy the pathogen that causes chickenpox **(1)**. So the pathogen that causes chickenpox could be able to grow fast enough to make you ill **(1)**.
2. If most children are immune to a disease as a result of vaccination **(1)**, the chance that an unvaccinated child will come into contact with an infected child is very small **(1)**.

6. Immunisation

1. (a) There is a very small risk that the child might suffer a serious illness as a result of having the vaccine **(1)**.
   (b) There is a much greater risk of serious illness if the child has the disease **(1)**.
2. Any suitable answer, for example the new cases might be caused by people catching the infection elsewhere and returning to the UK with it, or caused by a new strain of the virus that the vaccination didn't give immunity to **(1)**.

7. Treating diseases

1. **B** Overuse of antibiotics means only resistant bacteria survive **(1)**.
2. When the antibiotic is used, any non-resistant bacteria are killed, but individuals that are resistant as a result of a mutation survive and reproduce, which increases the proportion of resistant bacteria in the population **(1)**. If a new antibiotic is used against this strain, any individual bacteria that are resistant as a result of a mutation will survive and reproduce, so a population of bacteria resistant to the two antibiotics will increase in size **(1)**. This process occurs for each antibiotic used, which results in a strain of tuberculosis bacterium resistant to many antibiotics **(1)**. The concern is that there may be no new antibiotics that kill that strain so there will be no way of controlling the infection and it may cause an epidemic/pandemic **(1)**.

8. Cultures

1. blown in from the air **(1)**; transferred by touch **(1)**; in the culture media/on the equipment **(1)**
2. The Petri dish and the growth medium must be sterilised before use **(1)**. The inoculating loop must be sterilised by passing it through a flame and allowing it to cool **(1)**. The loop is then dipped into the bacterial culture and used to inoculate the growth medium in the Petri dish, keeping the dish as well covered as possible to reduce contamination **(1)**. After the disinfectant is added, the lid of the dish is secured with adhesive tape to prevent contamination from the air **(1)**.

9. Biology six mark question 1
Answers can be found on page 107.

10. Receptors

1. (a) stimulus: touching sharp pin **(1)**; response: moving hand away **(1)**
   (b) pain receptors in the skin **(1)**
2. (a) A stimulus is a change in the environment that produces a response in an organism **(1)**.
   (b) A receptor is a specialised cell that responds to a particular stimulus/change in the environment **(1)**.
   (c) The response is the change in the organism that is caused by a particular stimulus **(1)**.
   (d) A sense organ is an organ that contains specialised receptor cells **(1)**.

11. Responses

1. The reflex arc consists of a sensory neurone that carries the electrical impulse from the receptor to the central nervous system **(1)**. The impulse passes to a relay neurone which carries the impulse to a motor neurone **(1)**. The motor neurone carries the impulse to the effector **(1)**.

2. Simple reflex actions only involve a few neurones and this makes them rapid (**1**); they are also rapid because they occur without thought/are automatic (**1**). Simple reflex actions are important for survival or to protect us from harm, such as withdrawing from pain or blinking when something is near the eye (**1**).

12. Controlling internal conditions

1. One possible answer is: hormone FSH (**1**), secreted by pituitary gland (**1**), target organ the ovaries (**1**).

2. If body temperature rises too high it will affect the way enzymes in the body work, (**1**) which could harm body processes and so make us ill (**1**).

13. The menstrual cycle

1. Progesterone and oestrogen inhibit/prevent the release of FSH by the pituitary (**1**). If there is no FSH, then no egg matures in the ovary (**1**). So no egg is released that could result in pregnancy (**1**).

2. FSH/follicle stimulating hormone from the pituitary gland causes eggs in the ovaries to mature and stimulates the release of oestrogen from the ovaries (**1**). LH/luteinising hormone stimulates the release of a mature egg from an ovary (**1**). Oestrogen inhibits the production of FSH so that no more eggs mature in that cycle (**1**).

14. Increasing fertility

1. The woman is given fertility drugs that contain FSH, which makes eggs mature in her ovaries (**1**). The drugs also contain LH, which stimulates mature eggs to be released (**1**). The released eggs are collected and fertilised with sperm outside her body (**1**). When the embryos have developed into tiny balls of cells, one or two are placed in the woman's womb to develop until birth (**1**).

2. Contraceptive pills are very effective at preventing pregnancy, so that a couple can decide when they want to have children (**1**). The pill has some positive side effects, such as reducing the risk of some cancers, and some negative effects, such as increasing the risk of thrombosis/blood clot (**1**). As long as a woman doesn't smoke heavily and is not greatly overweight, then the benefit of controlling fertility outweighs the risk of side effects (**1**).

15. Plant responses

1. Auxin stimulates shoot cells to elongate (**1**). Auxin inhibits (reduces) the elongation of root cells (**1**).

2. Gravity causes auxin to move to the lower side of the shoot (**1**). So cells on the lower side of the shoot are inhibited in elongation/don't elongate as much compared with cells on the upper side (**1**). So the cells on the upper side grow longer than those on the lower side and the root curves downwards (**1**).

3. The shoot would continue to grow straight up (**1**), because all the cells in the zone of elongation further down the shoot would receive the same amount of auxin (**1**) and so grow equally (**1**).

16. Plant hormones

1. Rooting powder contains plant hormones (**1**). The hormones stimulate the end of the cutting to develop roots (**1**).

2. to kill the weeds (**1**), so that the crops don't have to compete with them and can grow faster (**1**)

3. The use of selective weedkillers will reduce the amount of weeds in the crop that produce seeds for the birds so fewer birds will survive in the area (**1**). If we want to increase crop yield but reduce the harm to birds, we need to make sure that there is other food for the birds, such as by leaving some areas with weeds in (**1**).

17. New drugs

1. to make sure they are safe/not toxic (**1**); to make sure they work as expected/ test their efficacy (**1**); to find the right dose for treatment (**1**)

2. The benefit of using animals as models is that the drugs can be tested for safety without harming humans (**1**). A problem with using animals as models for humans is that the drugs may not work in the same way in the animal as they would do in a human (**1**). (*Other problems may be acceptable, e.g. many people are against the use of animals for testing.*)

18. Thalidomide and statins

1. The drug was not tested on pregnant women (**1**), so the effect was not seen until after women had used it (**1**).

2. Recommend for no diabetes group (**1**); since large reduction in cardiovascular disease but no increased risk of diabetes (**1**). Do not recommend for diabetes group (**1**); since reduction in cardiovascular counteracted by significant increase in diabetes risk (**1**).

19. Recreational drugs

1. They are dependent on/addicted to the drug (**1**) so they will suffer distressing withdrawal symptoms if they stop taking it (**1**).

2. The study compared the use of cannabis with IQ at 38 for a large number of people. It also tried to control for other factors that might have an effect, so this study is likely to produce repeatable results (**1**). The relationship between amount taken and effect on IQ also suggests a valid relationship (**1**). However, this is only one study, and other studies would need to be considered to check whether other factors might also be having an effect/whether the relationship between cannabis and IQ decrease was reliable (**1**).

20. Drugs and health

1. (a) Because if used sensibly, it is safe to drink (**1**).
(b) If drunk in large quantities, it can damage the body (**1**).

2. There are a range of possible arguments and conclusions for this question, such as:

- Arguments for: (one from) making tobacco products illegal will mean they can't be sold in shops, so this will make it more difficult for people to get hold of them; young people won't be able to start smoking so easily and so won't smoke all their lives (**1**).

- Arguments against: (one from) people don't like to be told what they should and shouldn't do – it's their choice to smoke; making it illegal means that many people will just change to buying it illegally, which will create a large illegal market like there is for other illegal drugs, making it more difficult to control (**1**).

- Possible conclusions: (one from) making tobacco illegal could reduce the number of people who smoke and develop related diseases, so this will reduce costs of treating them; making tobacco illegal may reduce the number of people with smoking-related diseases, but it could also cause other problems such as for the police who would have to deal with the illegal trade (**1**).

21. Drugs in sport

1. Increased heart rate delivers more oxygen and food to muscles (**1**), so the muscles can release more energy, making the athlete faster or stronger in their sport (**1**).

2. Any one argument for making it illegal, such as: Not all athletes have access to the artificial drug, so that isn't fair; athletes may take the artificial drug in much larger amounts than the body produces and so it may cause harm (**1**). Any one argument against making it illegal, such as: Athletes can naturally boost EPO by high altitude training and not all athletes can do that, so what's the difference from not having access to the artificial form (1). Conclusion based on arguments, such as: It was a good decision because it could mean that athletes have more similar EPO levels in a race (1).

22. Biology six mark question 2

Answers can be found on page 108.

23. Competition

1. so that the plants don't compete (1) for an environmental factor such as light/water/nutrients (1)
2. Predators of milk snakes may confuse them with the poisonous coral snakes (1) and avoid them because they know that coral snakes are poisonous (1).

24. Adaptations

1. The fur is camouflage against predators (1), so needs to be white against the snow in winter and brown against the ground and plants in summer (1).
2. Deep in the ocean there is no light, so the light patterns can help other organisms identify which species it is (1). This could be used for attracting a mate or for scaring away the other animals (1).

25. Indicators

1. Some species of invertebrates are only found in unpolluted water and others only in highly polluted water (1). So by looking at what species are in the water you can see how polluted it is (1).
2. oxygen level (1), which is high in unpolluted water and low in polluted water (1)
3. Living organisms that are sensitive to pollution will die if there is a sudden increase in pollution (1), but otherwise the presence of an indicator organism usually shows the long-term level of pollution in that habitat (1). Non-living methods of monitoring measure the level of one pollutant at a particular time and need to be carried out on a regular basis to get a long-term record (1).

26. Energy and biomass

1. Some of the materials in the biomass will be lost to the environment in the rabbit's waste materials/faeces (1). Some of the biomass will be broken down in respiration and released as carbon dioxide (1).
2. A strength of the model is that it shows the loss of biomass as you go through a food chain (1); this helps us to understand that some biomass and energy are transferred to the environment at each level (1). A weakness of the model is that feeding relationships are not as simple as in the food chain described by the pyramid (1). The pyramid does not describe a food web or include decomposer organisms (1).

27. Decay

1. Warmth from the Sun will increase the rate of growth of microorganisms in the heap so they break the materials in the heap down faster (1).
2. Each crop that is grown in the ground takes nutrients out of the soil, leaving fewer nutrients for the next crop (1). Fewer nutrients means the plants won't grow as much, so they will produce a smaller yield (1).
3. Making compost with the waste reduces the amount of total waste sent to landfill, which is better for the environment (1). The compost can then be used by gardeners and farmers, instead of artificial fertilisers, to return nutrients to the ground where crops and plants are grown (1).

28. Carbon cycling

1. Microorganisms and detritus feeders break down the waste (1). The microorganisms and detritus use some of the materials from the waste for respiration, which releases carbon dioxide into the air (1).
2. (a) The carbon in fossil fuels has come from the carbon compounds in huge amounts of dead animal or plant material (1).
   (b) Combustion of fossil fuels releases large amounts of carbon dioxide into the air (1), faster than photosynthesis by plants removes it (1).

29. Genes

1. genes (1), environment/conditions during development (1)
2. Identical twins have the same genes because they come from the same fertilised egg, so characteristics controlled by genes will be the same (1). Any characteristics affected by the environment may be different between the twins because different things may happen to them as they grow up (1).

30. Reproduction

1. (a) They will have a range of different shapes and flower colour depending on which were their parent plants (1) because seeds are produced after fertilisation/sexual reproduction (1).
   (b) She should take cuttings (1) as all the plants grown from cuttings of the same plant will be genetically identical and so produce the same kind of flowers (1).

31. Cloning

1. (a) Any one suitable answer such as: use pieces of an adult plant, produces clones of the adult plant (1). *Just saying they are both done with plants is not enough because that information is given in the question.*
   (b) Any one suitable answer such as: plant tissue culture uses only a few plant cells but cuttings use much larger pieces of stem, root or leaf (1).
2. One adult animal provides a body cell, from which the nucleus is removed and used in the process (1). An adult female provides the unfertilised egg cell from which the nucleus is removed and thrown away, and the rest of the cell used in the process (1). The developing embryo is then placed into the womb of another female to grow and develop (1).
3. Embryo transplanting produces clones from one fertilised cell, which means the farmer can select the best cow as well as the best bull to produce all the embryos (1). The quality of the surrogate mother cows doesn't matter because their genes are not involved (1). If the bull was mated with each of the cows instead, the calves would all be different and would have some genes from their mother, so might not be as good quality (1).

32. Genetic engineering

1. The gene for herbicide resistance is put into some plant cells (1). The cells are treated/grown by tissue culture so they develop into new plants (1).
2. (a) It makes it easier to see which mice have the human disease gene in their cells (1).
   (b) The gene for the disease is cut out of a human chromosome and joined to a glow gene from a jellyfish (1). The combined genes are mixed with early mouse embryos so that the genes are taken into the cell nuclei and joined to the chromosomes (1).

33. Issues with new science

1. If the weeds become herbicide resistant, the farmer won't be able to kill them with herbicide and so the weeds will compete with the crop and reduce crop yield (1). This means the farmer will get less money when he sells his crop (1).

ANSWERS

2. Weed plants provide food for a wide range of animals **(1)**; any one example of use as food, such as nectar for insects, leaves for caterpillars and other herbivorous animals and seeds for seed-eating birds **(1)**. These animals are food for other species that live in the same habitat **(1)**. So completely removing weeds will affect the whole food web and could lead to extinction of many species **(1)**.

34. Evolution
1. **(a)** It shows that cows and pigs are more similar to each other **(1)** than they are to humans **(1)**.
   (b) cow and pig **(1)** because the tree shows that common ancestor for cow and hedgehog lived further back in time than the common ancestor for cow and pig **(1)**
2. Using just a few characteristics to group organisms means you may get groups that are not really closely related **(1)**. Using more characteristics is more likely to produce a classification where the individuals in the group really are closely related **(1)**.

35. Theories of evolution
1. **(a)** Jack developed his ability to swim by training **(1)**. His daughter inherited his champion ability **(1)**.
   (b) Jack inherited characteristics from his parents that made it possible for him to become a champion swimmer by training **(1)**. His daughter inherited these characteristics from Jack and so could also become a champion swimmer by training **(1)**.
2. Many weed plants may have a mutation that would make them herbicide-resistant, but they would only become obvious where herbicide is used **(1)**. Also, the gene used in the GM crop for herbicide resistance may have been transferred from the crop plants to the weed plants by pollination **(1)**. The appearance of resistant weeds has happened rapidly as the herbicide is used on the GM crop, killing all non-resistant weeds and leaving space for resistant weeds to grow **(1)**.

36. Biology six mark question 3
Answers can be found on page 108.

Chemistry answers

37. Atoms and elements
1. **(a)** Sodium is made of only one sort of atom/it is made from sodium atoms only **(1)**.
   (b) Oxygen is a non-metal/oxygen is in a different group/oxygen is a different element **(1)**.
2. A – neutron **(1)**, B – electron **(1)**, C – nucleus **(1)**

38. Particles in atoms
1. **(a)** Protons = 19 **(1)**, neutrons = (40 − 19) = 21 **(1)**, electrons = 19 **(1)**
   (b) No **(1)**, because they have different numbers of protons/atomic numbers **(1)**.

39. Electronic structure
1. **(a)** 2,4 **(1)**
   (b) 2,8,6 **(1)**
   (c) 2,8,8,2 **(1)**
2. 13 crosses drawn **(1)**, correct number of crosses in each circle (2,8,3) **(1)**

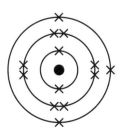

40. Electronic structure and groups
1. **(a)** Their highest occupied energy levels (outer shells) are full **(1)**. *It is not detailed enough to just to say that Group 0 has stable electronic structures.*
   (b) Their atoms all have one electron in their highest energy level/outer shell **(1)**.
2. 8 **(1)**; except for helium, which has 2 **(1)**
3. potassium + water → potassium hydroxide + hydrogen **(1)**; $2K + 2H_2O \rightarrow 2KOH + H_2$ **(1 mark for correct formulae, 1 mark for correct balancing)**

41. Making compounds
1. They gain electrons **(1)** to gain a full outer shell **(1)**.
2. A sodium atom loses one electron **(1)** to form a positive ion **(1)**.
3. **(a)** ions **(1)**
   (b) covalent **(1)**
4. Potassium atoms transfer electrons **(1)** to chlorine atoms **(1)** to form potassium ions and chloride ions **(1)**.

42. Chemical equations
1. (50 − 22) = 28 g **(1)** 28/2 = 14 g **(1)**
2. No atoms are made or lost in chemical reactions **(1)** and there are two H atoms and two F atoms on each side **(1)**
3. **(a)** $CaCO_3 + 2HCl \rightarrow CaCl_2 + H_2O + CO_2$ **(1)**
   (b) $2Na + Cl_2 \rightarrow 2NaCl$ **(1)**
   (c) $4Al + 3O_2 \rightarrow 2Al_2O_3$ **(1)**

43. Limestone
1. Any two for one mark each from: sand, aggregate, water **(2)**.
2. Any two advantages for one mark each: new jobs, makes money, provides building materials, makes useful products (or named product such as cement) **(2)**. Any two disadvantages for one mark each: fewer tourists, more traffic, noise, dust, visual pollution, named pollutant such as carbon dioxide (from the vehicles), damage to wildlife **(2)**.
3. **(a)** cost of fuel **(1)**
   (b) Carbon dioxide **(1)** causes global warming **(1)**, OR sulfur dioxide **(1)** causes acid rain **(1)**.

44. Calcium carbonate chemistry
1. **(a)** Sodium carbonate reacts with acids to produce carbon dioxide **(1)**, which escapes as bubbles of gas **(1)**.
   (b) Pass the bubbles through limewater **(1)**, which turns cloudy if the gas is carbon dioxide **(1)**.
2. calcium oxide + water → calcium hydroxide **(1)**
3. **(a)** Copper oxide **(1)**
   (b) $CuCO_3 + 2HCl \rightarrow CuCl_2 + CO_2$ **(1 mark for correct products in either order, 1 mark for correct balancing) (2)**

45. Chemistry six mark question 1
Answers can be found on page 108.

46. Extracting metals
1. Electrolysis of a molten potassium compound **(1)** because potassium is more reactive than carbon **(1)**.
2. **(a)** It needs a lot of energy **(1)** and many stages **(1)**.
   (b) Two points from the following, for one mark each: iron could be extracted using electrolysis/but carbon is cheap/electricity is expensive **(2)**.
3. $2ZnO + C \rightarrow 2Zn + CO_2$ **(1 mark for correct formulae, 1 mark for balancing)** *Alternative answer: $ZnO + C \rightarrow Zn + CO$*

47. Extracting copper
1. Supplies of high-grade copper ores are running out/only low-grade copper ores left in the future **(1)**; copper is extracted from copper ore so future supplies may be limited **(1)**.

103

ANSWERS

2. The negative electrode **(1)** because positive copper ions move to it **(1)**.
3. **(a)** $Cu_2S + 2O_2 \rightarrow 2CuO + SO_2$ **(1)**
   (b) Iron is more reactive than copper **(1)** so it displaces copper from its solutions **(1)**.
   (c) Two from the following for one mark each: less energy needed/plants release energy when they are burned/less waste produced/low-grade ores can be used/sulfur dioxide is not released **(2)**.

48. Recycling metals

1. Two from the following for one mark each: less waste rock produced/less carbon dioxide (or named pollutant) produced/less waste sent to landfill **(2)**.
2. Less energy is needed to melt aluminium than to melt steel **(1)**; more energy is needed to extract aluminium by electrolysis than is needed to extract iron in the blast furnace **(1)**.
3. One environmental impact for one mark, e.g. less quarrying (or specific impact of this such as less dust), less fuel used, less carbon dioxide produced, less landfill needed **(1)**. One ethical or social impact for one mark, e.g. local jobs created, resources saved, less noise pollution from local lorries or quarries **(1)**.

49. Steel and other alloys

1. A mixture of metals/a mixture of a metal and another element **(1)**.
2. Iron from the blast furnace contains impurities **(1)**, which make it brittle **(1)**.
3. Aircraft parts need to be light but strong **(1)**; duralumin is slightly denser than aluminium so parts will be heavier **(1)**, but duralumin is three times stronger **(1)** so is a better choice overall **(1)**.

50. Transition metals

1. Copper does not react with water **(1)** and it is a good conductor of heat **(1)**.
2. **(a)** low carbon steel **(1)**
   (b) Advantage: Aluminium is less dense than steel **(1)**, which makes the car lighter **(1)**; OR aluminium does not corrode **(1)**, which means that the car will last for longer **(1)**. Disadvantage: Aluminium is not as strong as steel **(1)**, which could mean the car is less safe in an accident **(1)**.
   (c) Identification of aluminium or titanium as better **(1)** but with reasons. For example, three from: aluminium is cheaper than titanium **(1)**, aluminium is less dense **(1)**, titanium is more expensive **(1)**, but titanium is 4 times stronger **(1)**, titanium is 1.7 times denser than aluminium but 4 times stronger **(1)**.

51. Hydrocarbons

1. **(a)** Contains only single bonds (between carbon atoms) **(1)**
   (b) Compound of hydrogen and carbon atoms only **(1)**
2. C_nH_{2n+2} **(1)**
3. The formula for X follows the general formula for alkanes (the number of hydrogen atoms is two plus twice the number of carbon atoms) **(1)**.
4.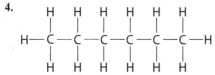
   Correct number of atoms **(1)** bonded correctly **(1)**

52. Crude oil and alkanes

1. They have different boiling points (or different ranges of boiling points) **(1)**.

2. Oil is evaporated **(1)** and passed into column, which is hot at bottom and cool at top **(1)**; the alkanes rise and condense at different heights/levels **(1)**.
3. Hydrocarbons with small molecules are less viscous (more runny) **(1)** and more flammable (easier to ignite) **(1)**.

53. Combustion

1. **(a)** Fuel A because hydrocarbons contain hydrogen and carbon **(1)** and these oxidise to water vapour and carbon dioxide **(1)**.
   (b) Fuel B **(1)** because during incomplete combustion carbon oxidises to form carbon monoxide **(1)**.
2. $C_3H_8 + 5O_2 \rightarrow 3CO_2 + 4H_2O$ (1 mark for correctly balancing each side) **(2)**

54. Biofuels

1. Carbon dioxide and water vapour are produced from ethanol **(1)**, but only water vapour from hydrogen **(1)**.
2. Less food grown/food prices increase **(1)**
3. $C_2H_5OH + 3O_2 \rightarrow 2CO_2 + 3H_2O$ (1 mark for correctly balancing each side) **(2)**

55. Chemistry six mark question 2

Answers can be found on page 108.

56. Cracking and alkenes

1. **(a)** Add bromine water and shake **(1)**; this turns from orange to colourless if alkenes are present **(1)**. *Writing 'clear' is not enough detail, as 'clear' just means see-through – it does not mean the same as 'colourless'.*
   (b) Position Y because it is a gas/it is made of small molecules **(1)**.
   (c) 1 mark for correct structure (check carefully that each C atom only has four lines attached to it).

   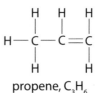

   propene, C_3H_6

   (d) to produce smaller hydrocarbons that are more useful as fuels **(1)** and alkenes, which are used to make polymers **(1)**

57. Making polymers

1. Any two from the following for one mark each: absorbs a lot of water, non-toxic (not poisonous), soft, cheap (because the nappy is disposable) **(2)**
2. **(a)**

   $$\left[\begin{array}{c} H \quad H \\ | \quad\; | \\ C - C \\ | \quad\; | \\ H \quad Cl \end{array} \right]_n$$ **(1)**

   (b) poly(chloroethene) (not 'poly(chloroethane)') **(1)**
3. Saturated **(1)**; because it only contains carbon–carbon single bonds (or no double bonds) **(1)**.

58. Polymer problems

1. Statement for first mark with reason for second mark. For example: polymers are made from crude oil **(1)**, which is a limited resource, so recycling conserves oil **(1)**; or recycling reduces the amount of polymer going to landfill **(1)** and we are running out of sites for landfill sites **(1)**.
2. **(a)** It is not a hydrocarbon **(1)** because it contains oxygen (or does not only contain hydrogen and carbon) **(1)**.
   (b) can be broken down or decomposed by microbes **(1)**
   (c) The bags are biodegradable/break down easily after use **(1)**.

59. Ethanol

1. Fermentation uses sugar from plants/a renewable resource **(1)**. If hydration of ethene was used, the raw material would be crude oil/a non-renewable resource **(1)**.

2. bubbles **(1)**

3.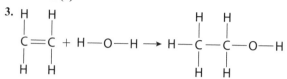

   1 mark for correct reactant structures, 1 mark for correct product structure **(2)**.

4. One advantage and one disadvantage of each method for 1 mark each. Advantages of fermentation: raw material is renewable, moderate temperatures needed **(1)**. Disadvantages of fermentation: slow reaction, product is impure, farmland is needed **(1)**. Advantages of hydration of steam: fast reaction, product is pure **(1)**. Disadvantages of hydration of steam: raw material is non-renewable, high temperatures needed **(1)**.

60. Vegetable oils

1. Any two from the following for one mark each: potato cooks faster in oil, potato cooked in oil has different flavours, boiled potato is soft **(2)**.

2. Water is heated to make steam **(1)**, the steam passes through the crushed plants and carries oil with it **(1)**, the vapours are condensed **(1)**, oil floats on water (and water is run off) **(1)**.

61. Emulsions

1. One mark for a property and one mark for a reason. For example, emulsions are thicker (than oil or water alone) **(1)** so they do not drip easily **(1)**; OR emulsions have good coating ability **(1)** so they stick to the wall well **(1)**.

2. (a) Oil does not dissolve in vinegar (which is mainly water) **(1)** and oil is less dense than water **(1)**.
   (b) Egg yolk contains an emulsifier **(1)** which stabilises the emulsion (stops the oil and vinegar separating) **(1)**.

3. (a) The emulsifier molecule has a hydrophilic 'head' **(1)** and a hydrophobic 'tail' **(1)**.
   (b) The hydrophilic part dissolves in the water droplets **(1)** and the hydrophobic part dissolves in the oil **(1)**, which stops the water droplets joining up again **(1)**.

62. Hardening plant oils

1. (a) unsaturated oil **(1)**
   (b) Bromine water changes from orange **(1)** to colourless **(1)**. *'Clear' is not the same as 'colourless' and is not correct.*

2. (a) Hydrogen is added **(1)** to the oil at 60°C with a nickel catalyst **(1)**, which adds hydrogen to the carbon=carbon double bond **(1)**.
   (b) Hardened vegetable oils have higher melting points **(1)** so they are solid at room temperature **(1)** and useful in margarine/cakes/pasties **(1)**.

63. The Earth's structure

1. a core at the centre **(1)** surrounded by the mantle **(1)** with a thin crust on top **(1)**

2. Convection currents in the mantle **(1)** driven by heat released by radioactive processes **(1)** move the plates.

3. (a) Any two from the following for one mark each: movement of tectonic plates; plates collide or push against each other; sudden movements happen **(2)**.
   (b) Any two from the following for one mark each: scientists do not know what is happening under the surface; they do not know where forces are building up; it is difficult to

measure these forces; they cannot know when the forces have reached their limit **(2)**.

64. Continental drift

1. The continents were once all joined together **(1)** and then moved apart **(1)**.

2. Any two from the following for one mark each: a land bridge could explain the common fossils; different species are on each continent today; Wegener could not explain how continents could move **(2)**.

3. 2 800 000 m/140 000 000 year **(1)** = 0.02 m/year (2 cm per year) **(1)**

65. The Earth's atmosphere

1. (a) $2Cu + O_2 \rightarrow 2CuO$ **(1)**
   (b) volume of oxygen = 100 − 79 = 21 cm³ **(1)**; percentage of oxygen = 21/100 × 100 = 21% **(1)**

2. (a) 200°C is below the freezing point of carbon dioxide **(1)** so the carbon dioxide would become a solid **(1)** (and contaminate the column or the gases).
   (b) The top of the column is above the boiling point of nitrogen **(1)** so it boils there and is given off as a gas **(1)**, but the bottom of the column is below the boiling point of oxygen **(1)** so it remains as a liquid **(1)**.

66. The early atmosphere and life

1. Any two from the following for one mark each: Earth has: more nitrogen, oxygen; less carbon dioxide **(2)**.

2. (a) Hydrocarbons (such as methane) interacted with ammonia and water **(1)** and lightning **(1)** to produce compounds needed for life/amino acids **(1)**.
   (b) Sensible suggestion, e.g. life formed a very long time ago, experiments cannot prove how life started **(1)**.

67. Evolution of the atmosphere

1. Photosynthesis by plants and algae **(1)** has reduced the percentage of carbon dioxide **(1)** and increased the percentage of oxygen **(1)**.

2. (a) It is too hot for water vapour to condense **(1)**.
   (b) The Earth's surface had to cool below 100°C **(1)**, and it took this long for it to happen **(1)**.
   (c) It took time for the oceans to form and dissolve carbon dioxide **(1)**, and photosynthesis could only begin when plants and algae had evolved **(1)**.

68. Carbon dioxide today

1. (a) They absorb large amounts of carbon dioxide from the atmosphere **(1)**.
   (b) increased use of fossil fuels (or a specific example of this, such as more cars) **(1)**

2. Human activities increase the amount of carbon dioxide in the atmosphere **(1)**. Example of such activities for one mark, e.g. more fossil fuels used (e.g. more cars or power stations); extra carbon dioxide cannot all be absorbed (e.g. fewer forests) **(1)**.

69. Chemistry six mark question 3
Answers can be found on page 109.

Physics answers

70. Infrared radiation

1. (a) The black tubes absorb infrared radiation (which is transferred to the water as heat energy) **(1)**.
   (b) Black is the better absorber and emitter of infrared light **(1)** so it will collect more radiation from the Sun and heat the air to a higher temperature **(1)**.

ANSWERS

71. Kinetic theory

1. Solid: The teacher holds the box still or hardly vibrating **(1)** so that the balls remain packed closely together **(1)**. Liquid: The teacher gently shakes the box **(1)** so that the balls move around the bottom of the box **(1)**. Gas: The teacher shakes the box vigorously **(1)** so that the balls move and hit all the sides of the box **(1)**.

72. Methods of transferring energy

1. Any two from: Energy transfers from your hand into the perfume **(1)**. As the temperature of the perfume rises, more of the particles in it evaporate **(1)**. The remaining particles in the perfume have less energy, so the liquid feels colder **(1)**.

2. Gold is a metal **(1)**, so has free electrons, whereas polythene does not **(1)**.

73. Rate of energy transfer

1. The drink molecules with the most energy evaporate from the surface **(1)**. Blowing the molecules away stops them from condensing back **(1)** into the drink, so the average energy/temperature of the drink molecules falls **(1)**. *You could also say that blowing pushes cooler air over the drink so increasing the rate of convection.*

2. In hot climates animals have to transfer energy from their bodies to maintain the right temperature **(1)**. Ears with large surface areas will encourage energy transfer by conduction and convection **(1)**. OR In cold climates animals must maintain their body temperature **(1)** so ears with small surface area reduce energy transfer by conduction and convection **(1)**. Hence the correlation is not chance.

74. Keeping warm

1. The filled cavity cuts the energy transferred through the walls by convection **(1)** because air is trapped in small pockets **(1)** so the wall becomes a better insulator **(1)**.

2. cost of energy lost through walls is 35% × £1200 = £420 **(1)**. Installing cavity wall insulation saves two thirds of this, i.e. 2/3 × £420 **(1)** = £280 **(1)**.

75. Specific heat capacity

(a) concrete energy = 100 kg × 960 J/kg°C × 60°C **(1)** = 5 760 000 J **(1)**; brick energy = 100 kg × 900 J/kg°C × 60°C **(1)** = 5 400 000 J **(1)**

(b) The brick requires less energy to raise its temperature **(1)** but transfers less energy to the room **(1)**; the concrete requires more energy to raise its temperature but transfers more energy to the room **(1)**, so it is more effective to use concrete **(1)**.

76. Energy and efficiency

1. (a) cost of energy saved each year = cost of oven/payback time **(1)** = £450/9 years **(1)** = £50/year **(1)**

(b) The new oven has a higher efficiency **(1)** so the energy input can be lower **(1)** to give the same useful energy output.

77. Physics six mark question 1

Answers can be found on page 109.

78. Electrical appliances

1. energy transferred = 2.2 W × 5 × 60 s **(1)** = 660 J **(1)**

2. (a) energy transferred = 0.04kW × 2 h = 0.08 kWh **(1)**; cost = 0.08 kWh × 12 p/kWh = 0.96p **(1)**

(b) time = energy transferred/power **(1)** = 0.08 kWh/0.0032kW **(1)** = 25 hours **(1)**

79. Choosing appliances

1. The temperature is kept below 6°C by A for 60 min **(1)** and by B for 100 min **(1)**. B is therefore more effective for keeping the medicine safe **(1)**.

2. Time = energy transferred/power **(1)**; time that each bike can run for is: Whizz = 0.52 kWh/0.25 kW = 2.08 h **(1)**; Bikee = 0.36 kWh/0.2 kW = 1.8 h **(1)**. The Whizz/more expensive is the most effective as the power will last for longer **(1)**. *You could also say there are not enough data to compare how far the bikes could travel in the time. If the cheaper bike is a lot lighter then it could travel further in the shorter time.*

80. Generating electricity

1. Any two from: Nuclear power stations are expensive to build **(1)** and to decommission **(1)** and they also require expensive safety measures **(1)**.

2. Any four points from: Reduce demand for electricity **(1)**, by encouraging energy saving measures **(1)**. Replace the old power stations with natural gas power stations because they are the cheapest **(1)**, or build new nuclear power stations because they have a long life and they produce electricity relatively cheaply **(1)** (each suggestion must be backed up by a reason), using a renewable energy source **(1)**.

81. Renewables

1. Tidal power is expensive to construct **(1)**; but it is regular/reliable because tides rise and fall twice a day **(1)**. Hydroelectric dams are expensive to build **(1)**; but lots of sites/can be used at any time/reliable as any rainwater can be stored **(1)**.

2. Wind farms are expensive to build, but there are no fuel costs, which means that the electricity is not very much more expensive with wind farms **(1)**. A rise in fuel costs could make offshore wind cheaper than coal **(1)**. Wind is not reliable **(1)** while coal can be used anytime that fuel is available **(1)**. *Other valid points include the difficulty/cost of maintaining turbines sited out at sea.*

82. Environment and energy

1. Advantage: Carbon dioxide is not released into the atmosphere (at the power station) **(1)**. Disadvantage: coal has to be mined and transported and this destroys habitats/creates noise/visual pollution/emissions from machinery and vehicles **(1)**.

2. Using anaerobic digesters stops the methane formed in landfill sites from being released into the atmosphere **(1)**. Using methane from waste to generate electricity reduces the need for fossil fuels/natural gas to be burned and so reduces the contribution to global warming **(1)**.

83. Distributing electricity

1. Step-up transformers increase the voltage from power station delivered to the cables **(1)**. Step-down transformers reduce the voltage supplied to homes/consumers **(1)**.

2. (a) The cables lose less energy to the surroundings **(1)** because the current is very small **(1)**.

(b) Any two points from: Pylons are a source of visual pollution **(1)**, which may damage tourism **(1)** (or any other valid environmental problem caused by overground cables). The cost of this may outweigh the extra cost of burying the cables **(1)**. Burying cables may disturb habitats **(1)** and add to the cost of electricity **(1)**.

84. Physics six mark question 2

Answers can be found on page 109.

85. Properties of waves

1. **(a)** The ripples are transverse waves **(1)** because the particles of water move up and down/at right angles to the motion of the wave **(1)**, so only energy is carried to the edge of the bowl.
   (b) $f = 2$ Hz **(1)**; $\lambda = 16$ cm/4 = 4 cm **(1)**; $v = 2$ Hz $\times$ 4 cm **(1)** = 8 cm/s **(1)**

86. Electromagnetic waves

1. Any three points from: It is unethical because it uses children **(1)**. It is unscientific because there are no controls/comparison with children without mobile phones **(1)**. Possibly biased in favour of the phone company **(1)**. It may take a long time for any harm to show, and we don't know how long the study lasted **(1)**. 100 is not a very big sample size **(1)**.

87. Waves

1. Reflected waves do not cross the boundary but refracted waves do **(1)**. Reflected waves and refracted waves change direction **(1)**.
2. **(a)** Some of the infrared waves reach the TV because they are reflected **(1)** by the wall and change direction **(1)**.
   (b) The straw looks bent because light waves from under the water are refracted when they enter the air **(1)** and change direction **(1)**.
   (c) Sound waves are diffracted around the doorway but light waves are not **(1)** because the wavelength of sound waves is similar to the width of the door but the wavelength of light is much shorter **(1)**.
   (d) X-rays are diffracted by atoms in the crystal and light waves are not **(1)** because X-rays have a wavelength that is similar to the size of the atoms **(1)** but light has a longer wavelength **(1)**.

88. Reflection in mirrors

1. a diagram with a mirror opposite the driveway at an angle of about 45° **(1)**; rays from a car on the road reflected off the mirror to the driver in the driveway **(1)**; with the angle of incidence equal to the angle of reflection **(1)**

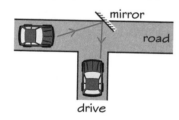

2. ray of light drawn from back of head to first mirror to second mirror to eye **(1)**; normals drawn **(1)**; angle I = angle r at each mirror **(1)**

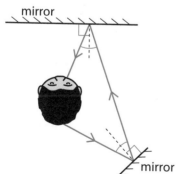

89. Sound

1. They communicate using sound **(1)**, which is longitudinal waves **(1)** that travel through the water.
2. The echo always takes a longer path than the original sound **(1)**, so its travel time is longer **(1)**.

3. diagram of a wave with crests and troughs further apart (greater amplitude) **(1)** and closer together (higher frequency) **(1)** than both A and B

90. Red-shift

1. Light from the more distant galaxies is red-shifted **(1)** more than **(1)** light from the galaxies that are closer.
2. The microwaves from the gun reflect off moving cars **(1)**. If the car is moving away the reflected waves have a lower frequency/longer wavelength/are red-shifted **(1)** (or the opposite for approaching cars) because of the Doppler effect. The change in frequency is related to the speed of the car **(1)**.

91. The expanding universe

1. Observations of red shift **(1)** show that galaxies are moving apart as predicted by Big Bang theory **(1)**. The cosmic microwave background radiation/CMBR **(1)** fills the universe/comes from every direction in space as predicted by Big Bang theory **(1)**.
2. The Big Bang theory is the only theory that predicts the CMBR (the microwave radiation left over from the Big Bang) **(1)** as well as the expansion of the universe **(1)**.

92. Physics six mark question 3

Answers can be found on page 109.

Six mark question answers

'Now try this' answer guidance:

- A basic answer is usually badly organized, has only basic information in it, does not use scientific words and includes poor spelling, punctuation and grammar.
- A good answer usually contains accurate information and shows a clear understanding of the subject. The answer will have some structure and the candidate will have tried to use some scientific words but it might not always be accurate and there may not be all the detail needed to answer the question. There will be a few errors with spelling, punctuation and grammar.
- An excellent answer contains accurate information, is detailed and is supported by relevant examples. The answer will be well organized and will contain lots of relevant scientific words that are used in the correct way. The spelling, punctuation and grammar will be almost faultless.

9. Biology six mark question 1

A basic answer: A brief explanation of one way that diet and exercise keep you healthy (for example, the benefits of having a balanced diet or why it is important to exercise). (Some credit might be given for answers that include general information on how to diet and exercise.)

A good answer: Some explanation of at least two ways that diet and exercise keep you healthy. For example, the answer might contain information on malnutrition and how it could lead to deficiency diseases or weight change. It might also include information on the benefits of exercise.

An excellent answer: There is a clear, balanced and detailed explanation of the nutrients found in a balanced diet, malnutrition, weight change and deficiency diseases.

Examples of points made in the response:

- A balanced diet provides all the nutrients you need in the right proportions, and so avoids malnutrition.
- These nutrients include carbohydrates, fats, proteins, also small amounts of vitamins and minerals.
- Malnutrition can lead to health problems such as deficiency diseases and Type 2 diabetes.
- The right amount of energy in the diet, which balances energy expended, will stop you becoming overweight or underweight.
- People who exercise regularly are more likely to stay healthy than people who don't exercise much. For example, it can help to prevent being overweight, which can increase the risk of Type 2 diabetes. (Other examples of ways in which exercise can affect health could be given as long as there is a clear link to how it improves health.)

22. Biology six mark question 2

A basic answer: A brief explanation of either how IVF is carried out or why multiple births carry more risk, but with little detail. (There might be a general reference to the way that IVF is carried out.)

A good answer: Some explanation of the process of IVF with some detail about the procedure and a comment about relative risks of multiple births. (Some credit might be awarded for general comments about the process of IVF.)

An excellent answer: There is a clear, balanced, well-ordered and detailed explanation of the process of IVF as well as analysis of the data provided by the graph and the two points are linked with a clear explanation.

Examples of points made in the response:

- In IVF, FSH and LH are used in fertility drugs to stimulate eggs to mature in the ovaries and to stimulate their release so that the eggs can be collected.
- Eggs are fertilised outside the mother's body and develop into embryos.
- One or two early embryos/at the small ball of cells stage may be placed in the mother's womb/uterus to develop.
- Placing two embryos in the mother's womb increases the chance that the couple will have at least one baby.
- The graph supports the advice.
- If IVF clinics place only one embryo in the womb/uterus, there is a better chance of survival of the baby after birth.
- The graph shows that twins have a greater risk of dying both after birth (approx. 17%) and within their first year (approx. 25%) than single babies (approx. 5% after birth, and approx. 6% in first year).
- Also babies born as triplets have an even greater risk of dying than those who are twins (approx. 29% after birth, approx. 48% in first year).

36. Biology six mark question 3

A basic answer: A brief description of one or two points given below, for example, plants take in carbon dioxide or all organisms give out carbon dioxide. (Some credit might be given for answers that include general information on respiration or photosynthesis.)

A good answer: Some explanation of the two main areas of the carbon cycle, such as the effect of photosynthesis and respiration on atmospheric carbon dioxide, and how carbon is passed along the food chain in biomass.

An excellent answer: There is a clear, balanced and detailed explanation of the role of organisms such as plants and algae, animals, microorganisms and detritus feeders, and the contribution of photosynthesis and respiration to the carbon cycle.

Examples of points made in the response:

- Respiration of all organisms (plants, algae, animals, microorganisms and detritus feeders) breaks down

carbon compounds and releases carbon dioxide into the atmosphere.

- Photosynthesis by green plants and algae takes carbon dioxide from the atmosphere.
- Inside plants and algae the products of photosynthesis are converted to carbohydrates, proteins and fats.
- When organisms die, some animals and microorganisms feed on their bodies.
- When animals, microorganisms and detritus feeders digest and absorb their food, some of the carbon compounds are used to make more carbohydrates, proteins and fats in their bodies.

45. Chemistry six mark question 1

A basic answer: A brief description of either an advantage or a disadvantage of using limestone as a building material. Little information from the table is used.

A good answer: A clear description of both an advantage and a disadvantage of using limestone as a building material. Information from the table is used in support.

An excellent answer: A clear, balanced and detailed description of advantages and disadvantages of limestone as a building material, fully supported using relevant knowledge and information from the table.

Examples of points made in the response:

Advantages:

- Less energy is needed to extract limestone than brick.
- Less air pollution is released (or named pollutant such as carbon dioxide).
- Reduced effect on the environment from air pollution (such as global warming).
- Natural appearance.
- Easily cut to different shapes.

Disadvantages:

- Limestone is more expensive than brick.
- Limestone has less resistance to air pollution.
- Limestone has to be quarried.
- Example of problems caused by quarries (such as noise, dust, extra traffic, destruction of habitats).

55. Chemistry six mark question 2

A basic answer: A simple description including one statement about the extraction of aluminium *or* copper.

A good answer: A clear description that includes either two statements about extracting aluminium *or* copper, or one statement about the extraction of copper *and* aluminium.

An excellent answer: A clear, balanced and detailed description of the ways that both aluminium *and* copper are extracted from their ores.

Examples of points made in the response:

Aluminium extraction:

- Aluminium cannot be extracted from aluminium oxide using carbon.
- Aluminium must be extracted using electricity (electrolysis).
- Electricity is expensive.

Copper extraction:

- Copper is extracted from its ore by heating in a furnace.
- The copper is purified by electrolysis.
- Copper can also be extracted from solutions of copper salts by displacement using scrap iron.
- Copper can be extracted by phytomining.
- Copper can be extracted by bioleaching.

69. Chemistry six mark question 3

A basic answer: A brief description is given of how the plates move *or* at least one result of their movement is given.

A good answer: A clear description is given of the effects of the movements of tectonic plates *or* a clear explanation is given of how they move, *or* a brief explanation is given of the effects *and* a brief description is given of how the plates move.

An excellent answer: A clear, detailed and balanced description is given of how tectonic plates move *and* an explanation is also given of how they move.

Examples of points made in the response:

How tectonic plates move:

*   Convection currents in the mantle move the plates.
*   The mantle is mostly solid but can move.
*   Currents are driven by heat from natural radioactive processes.
*   Plates move at a few centimetres per year.

The effects of tectonic plates moving:

*   Continents move apart or together.
*   Mountain-building happens when plates move together.
*   Plates also move against each other.
*   Sudden movements cause earthquakes.
*   Earthquakes can be disastrous.
*   Volcanoes can also result from the movement of plates.

77. Physics six mark question 1

A basic answer: A brief description of an advantage or a disadvantage of installing double-glazing.

A good answer: Either a brief description of an advantage and a disadvantage of installing double glazing or a more detailed description of an advantage or a disadvantage of installing double glazing, including information about cost or payback time.

An excellent answer: A clear, balanced and detailed description of both the advantages and disadvantages of installing double glazing with supporting evidence from the information provided.

Examples of points made in the response:

Advantages:

*   Double glazing (2.0) has a much lower U-value than single glazing (5.0).
*   This means that the rate of heat transfer through a double glazed window is much lower.
*   Less energy is lost from the house so the cost of energy is reduced (by £250 per year).
*   Cost of replacing windows with double glazing is recovered in 20 years.

Disadvantages:

*   Replacing windows with double glazing is expensive (£5000 instead of £2000).
*   The payback time is the cost/savings per year (£5000/£250 per year).
*   The payback time is long (20 years).
*   Even double glazed windows have a high rate of heat transfer – almost as much as a cavity brick wall (1.8).

84. Physics six mark question 2

A basic answer: A brief statement about the impact of either fossil fuels or renewable sources of energy on the environment.

A good answer: A brief statement about the effects of fossil fuels and renewable sources of energy on the environment or a more detailed statement about the effect of fossil fuels or renewable sources of energy on the environment.

An excellent answer: A clear, balanced and detailed description of the impact that both fossil fuels and renewable sources of energy have on the environment with some examples of renewable sources of energy given.

Examples of points made in the response:

Fossil fuels:

*   Fossil fuels release carbon dioxide into the atmosphere, which is linked to global warming.
*   Fossil fuels also release acidic gases, which can cause acid rain.
*   Fossil fuels produce waste materials such as ash.
*   Fossil fuel power stations can produce visual pollution.
*   Fossil fuels have to be extracted from the Earth and this can cause disruption of wildlife habitats (for example, oil spills).

Renewable sources of energy:

*   Renewable sources of energy produce less carbon dioxide.
*   Renewable sources of energy produce fewer waste materials.
*   Renewable sources of energy can cause visual and noise pollution (for example, wind turbines may be very obvious in a landscape).
*   Renewable sources of energy can cause destruction of wildlife habitats (for example, flooding of valleys to build hydroelectric power stations).
*   Wind turbines can kill birds.
*   Solar power farms cover a lot of land and destroy natural habitats.
*   Growing biofuels takes up a lot of land and destroys natural habitats.
*   Burning biofuels produces pollutants/waste materials.

92. Physics six mark question 3

A basic answer: A brief description of the Big Bang theory or a piece of evidence for the theory.

A good answer: Either a brief description of the Big Bang theory and a brief discussion of how either red-shift or the CMBR provide evidence for it, or a more detailed description of Big Bang theory or the evidence for it.

An excellent answer: A clear, balanced and detailed account of the Big bang theory and of both red-shift and the CMBR.

Examples of points made in the response.

*   The Big Bang suggests that the universe expanded from a point with a burst of energy.
*   The theory predicts that the universe has continued to expand.
*   Observations of distant galaxies show that they are moving apart.
*   And more distant galaxies are moving away faster than closer ones.
*   This shows the universe is expanding.
*   The evidence is from measurements of the red-shift of light received from the galaxies.
*   When light is emitted from an object moving away the wavelength increases.
*   The red-shift increases the further a galaxy is from us.
*   The Big Bang theory predicts that energy released when the universe began is still present today.
*   This background radiation fills space/cosmos.
*   It is called cosmic microwave background radiation.
*   The CMBR has been detected and investigated and agrees with the predictions of the Big Bang theory.
*   Only the Big Bang theory explains the red-shift of the galaxies and the CMBR.

Published by Pearson Education Limited, Edinburgh Gate, Harlow, Essex, CM20 2JE.

www.pearsonschoolsandfecolleges.co.uk

Copies of official specifications for all AQA qualifications may be found on the AQA website: www.aqa.org.uk

Text and original illustrations © Pearson Education Limited 2013
Edited by Judith Head and Florence Production Ltd
Typeset and illustrated by Tech-Set Ltd, Gateshead
Cover illustration by Miriam Sturdee

The rights of Peter Ellis, Sue Kearsey and Nigel Saunders to be identified as authors of this work have been asserted by them in accordance with the Copyright, Designs and Patents Act 1988.

First published 2013

16 15 14 13
10 9 8 7 6 5 4 3 2 1

British Library Cataloguing in Publication Data
A catalogue record for this book is available from the British Library

ISBN 978 1 447 94214 6

Printed in Slovakia by Neografia

Acknowledgements

We are grateful to the following for permission to reproduce copyright material:

Figures
Figure 14/B1.2.5 "Fresh and Frozen IVF Cycles, Live Birth rate by age", *HFEA publication Fertility Facts and Figures 2008*, p.6, http://www.hfea.gov.uk/docs/2010-12-08_Fertility_Facts_and_Figures_2008_Publication_PDF.PDF, copyright © HFEA.

Tables
Table 18/B1.3.2 "Cardiovascular benefits and diabetes risks of statin therapy in primary prevention: an analysis from the JUPITER trial" by Paul M Ridker, Aruna Pradhan, Jean G MacFadyen, Peter Libby, and Robert J Glynn, *The Lancet,* Volume 380, Issue 9841, pp.565–571, 11 August 2012, copyright © 2012, Elsevier.

In some instances we have been unable to trace the owners of copyright material, and we would appreciate any information that would enable us to do so.

All other images © Pearson Education

Every effort has been made to contact copyright holders of material reproduced in this book. Any omissions will be rectified in subsequent printings if notice is given to the publishers.

In the writing of this book, no AQA examiners authored sections relevant to examination papers for which they have responsibility